Plastics in the Oceans

Increasing plastic pollution, particularly in oceans, calls for a fundamental shift toward future-proof plastics throughout the circular economy. In a landmark move in 2022, the United Nations adopted a historic resolution to craft an international legally binding instrument to tackle plastic pollution. In line with this global initiative, *Plastics in the Oceans* investigates the problem of plastic pollution in the oceans, proposes solutions relating to current and advanced materials and technologies, and offers an optimal sustainable strategy to eliminate plastic pollution by 2040. The title:

- Addresses the entire life cycle of plastic, from production and design to disposal, and advocates for a comprehensive approach to mitigate plastic pollution.
- Focuses on primary and secondary microplastics, recognizing their significant impact on marine ecosystems.
- Explores various alternative materials, including biobased biodegradable polymers, as viable alternatives to conventional plastics.
- Provides insights into regulatory frameworks pertaining to plastic pollution across major world regions.
- Elucidates the need for reducing plastic production and enhancing recycling efforts in combating plastic pollution.
- Suggests strategies aligned with the principles of the 3R's—reduce, redesign, recycle—to foster the development of sustainable and future-proof plastics.

This book is aimed at technical readers in environmental engineering, chemical engineering, and polymer chemistry, as well as general readers interested in scientific solutions to the plastics pollution problem.

Plastics in the Oceans
Toward Sustainable Solutions

Jean-Luc Wertz, Olivier Bédué, and
Serge Perez

CRC Press
Taylor & Francis Group
Boca Raton London New York

CRC Press is an imprint of the
Taylor & Francis Group, an **informa** business

Designed cover image: Shutterstock

First edition published 2025
by CRC Press
4 Park Square, Milton Park, Abingdon, Oxon, OX14 4RN

and by CRC Press
2385 NW Executive Center Drive, Suite 320, Boca Raton FL 33431

CRC Press is an imprint of Informa UK Limited

ISBN: 978-1-032-86996-4 (hbk)
ISBN: 978-1-032-87395-4 (pbk)
ISBN: 978-1-003-53247-7 (ebk)

DOI: 10.1201/9781003532477

Typeset in Times
by Apex CoVantage, LLC

Contents

Foreword 1

Plastics in the Oceans: Toward Sustainable Solutions, the title of the book from Jean-Luc Wertz, Olivier Bédué, and Serge Perez, could appear as a kind of cliché in this early 21st century. Therefore, it is necessary to clarify what that means.

It depends partly on the reader: for the scientist, "plastic" means a material based on giant molecules (macromolecules), while it means wasted lightweight objects for the average person. This last definition is clearly the one the authors wanted to deal with when writing their book. Moreover, a more subtle distinction has to be made between objects ("macroplastics") and particles ("microplastics"), which are almost invisible. All themes are treated in this review.

Everyone knows what "oceans" means, right?

On the whole Earth, there are about 1.4 billion, billion tons (1.4×10^{18} tons) of water in the oceans, of which about 3×10^{17} tons (20%) are at less than 1000 m in depth and thus close enough from the surface to receive a fraction of sunlight.

Compared to these amounts of water, the approximate amount of plastics in the oceans, about 200 million tons, represents less than either one- or ten-thousandths of a part per billion!

Regardless, as the book demonstrates, such a seemingly "homeopathic" level can be of vital importance to all life—including human life—on Earth.

Are we human beings so carefree that we are willing to put our own lives at risk throwing away so much plastic? Yes and no! Yes, because everyone sees how many so-called single-use objects are wasted daily. But no, as we do not see the huge amounts of "primary microplastics" entering the oceans: only particles of abraded tires (while driving) and microfibers from abraded synthetic textiles (during laundry) account for around a million tons dumped into the oceans each year.

But are plastics—macroplastics or microplastics—really so toxic?

Pictures of turtles that died after eating single-use plastic bags instead of jellyfish have been shared around the world. But no human being will eat any macroplastic, and even so, if small pieces of plastic are eaten, as plastics are water-insoluble, they will be naturally eliminated. But what about micro- or even nano-particles? They can cross the body's biological barriers and enter the bloodstream and the tissues of various organs. Such a reality, explained in the book, should not only scare us but make us aware of the need for solutions to the problem of plastic pollution at every level.

And solutions for combating plastic pollution are well proposed in the third chapter of the book. From the 5Rs rule (refuse, reduce, reuse, repurpose, and then recycle) to the chemical syntheses of new materials bio- or oxo-degradable, the coverage of solutions appears quite comprehensive.

At the end of the work, behavioral and even philosophical considerations are put forward. Indeed, the visible part of plastics in the oceans—only a part floats on the surface—represents a very small aspect of the general problem of the reality of terrestrial pollution.

More dangerous are the pollutions by various water-soluble chemicals arising from human activities and dispersed in the sea at molecular levels. Combating this general problem is fighting against the universal law of entropy.

From this viewpoint, single-use plastic floating on the surface of the oceans is a kind of indication of the level of general pollution of the Earth. In case of illness and fever, the thermometer is useful: breaking it does not solve the problem! Likewise, plastic floating in the oceans is only one indicator of a level of pollution. Eliminating it is not the only solution; as Czigany[1] said: a strategy is needed toward "a radical reduction in throwing away unwanted and unusable products."

Jacques Devaux, Emeritus Professor
Université catholique de Louvain
Louvain-la-Neuve, Belgium

NOTE

1   T. Czigany, 2020, *Polymer Letters*, Vol. 14, No.1, https://doi.org/10.3144/expresspolymlett. 2020.1

Foreword 2

Mare Nostrum. By thus designating the Mediterranean Sea, the Romans showed their domination over the Mediterranean shores and on this sea, which, in facilitating trade and cultural exchanges, contributed to the strength of the Roman Empire. This expression could be extended to the world's seven seas, emphasizing our dependence on the oceans and our responsibility to preserve their integrity.

Marine pollution has become a growing concern in this century. The majority of the pollutants that are found in the oceans come from human activities. Manufactured products littering the seas are essentially plastic debris. Such wastes are particularly problematic since their decomposition can take hundreds of years. Of the 350 million metric tons of plastic waste produced each year, 1.7 million metric tons (i.e., 0.5% of the world's plastic waste) end up in the oceans. This constitutes around 80% of all marine debris found from surface waters to deep seas, and this is projected to triple in the next 20 years, adding 23–37 million metric tons of waste into the ocean per year.

The book analyzes the sources of plastics, their industrial transformation, and the complex mechanisms by which they are converted through fragmentation processes, chemical alteration of their structure, and biochemical modification. The resulting degraded products will, for one part, be further metabolized by marine organisms, and for the unmetabolizable part, be aggregated into massive floating garbage. The Great Pacific Garbage Patch is a striking example of the urgent need for solutions to cleanse plastic accumulation from the oceans. Due to their longevity and resistance to decomposition, resilient buoyant plastic wastes can be transported over extended distances where they persist and accumulate in patches. The Great Pacific Garbage Patch has been estimated to cover more than 1.5 million square kilometers of surface area. Several solutions to the circular economy are being proposed. Architects are devising revolutionary concepts: an example is the futurist design "Polimeropolis," an architect's visualization of a floating archipelago harboring a model city, essentially built from recycled floating plastics and allowing the capture of micro*plastics*.

Evaluations of plastic sources and sinks have often given inadequate representation of plastic size, yet size influences their distribution between surface and deep waters.

Tackling ocean pollution by plastics requires an in-depth scientific understanding of various plastics' chemical and physical characteristics. Such an approach is typically the domain of highly specialized scientists with long expertise in the various aspects of plastic chemistry and bioconversion.

The three authors, Jean-Luc Wertz, Olivier Bédué, and Serge Perez, all renowned specialists internationally recognized for their contribution to the scientific domains of polymers and plastics, brought together their complementary expertise to produce the in-depth analysis of the problems of "plastics in the oceans." They cover polymer physical chemistry, chemical engineering, structural and conformational analysis, computational chemistry, and molecular modeling. In addition to the purely chemical and physical domains, they add considerations in economic science.

The present treatise, *Plastics in the Oceans: Toward Sustainable Solutions*, although it deals with the pollution of the marine environment, is not, strictly speaking, a book covering all sources and all aspects of ocean pollution. This book distinguishes itself from the other books treating marine pollution from a more general point of view in that its scope focuses exclusively on plastics, from synthesis, uses in industry, and daily life, to their rejection in the marine environment. On the basis of their macromolecular characteristics and chemical structures, the fate of plastics and their respective biodegradability are examined from land to oceans. Based on these considerations, and in keeping with the increasing demand for the production and consumption of plastic polymers, ways to balance recycling and renewability are investigated, analyzed, and discussed.

Plastic pollutants in the oceans are of two categories: those that are degradable and disappear with time and those that remain or are broken down into nondegradable elements and nano-elements. The complex mechanisms of plastic degradation involve a chain process initiated by photooxidation, leading to fragmentation. The embrittled plastic surface then becomes susceptible to microbial enzymes and undergoes biodegradation and biochemical modifications that end up with micro- and nano-plastics entering the food chain.

The book is structured into four chapters, each subdivided into numerous complementary and detailed subsections. Each chapter is associated with a large and accurate bibliography. The authors provide a comprehensive survey of the problems of plastics in marine pollution. Thus, Chapter 1 presents a clear introduction to the gravity of the pollution engendered by plastic biodegradation, becoming worse due to the ever-increasing amounts of global production. It also presents the fate of plastic in the ocean environment with the various forms of biodeterioration and biodegradation. In this first chapter, the global market and the regulations aiming at reducing fossil resources consumption are evaluated, and strategies to convert it to produce fossil plastic are envisaged. Chapter 2 inventories the sources of marine plastics, their life cycle, their ecological and biological impact, their social and economic implications, and the regulations recommended by global and national institutions of governance. The responsibility of stakeholders and producers is underscored. The third chapter is dedicated to the examination of the solutions envisaged to mitigate and control marine pollution by plastics. A rich review and discussion of possible solutions toward a circular global economy is presented. A large part is devoted to biodegradability, compostability, and biobased polymers. The proposed goals and international agreements are reviewed. The various recycling technologies, important managing strategies to reduce plastic pollution and part of the "3R's principle" defined by the authors in their conclusions, are described and discussed in detail. The last parts of this important chapter can be regarded as a type of large debate on the respective interests of fossil-based biodegradable polymers versus polymers from natural origin. Chapter 4 brings a conclusion, here again, subdivided into several sections in which the authors consider the different attitudes and policies of the leading countries in their efforts to harness ocean pollution by plastics. Regulations taken in each country by the decision-makers are presented. The authors deliver their own views and opinions on the necessary research and development that needs to be implemented. Rather than envisaging a near-complete eradication of fossil plastics, the authors advocate biobased polymers and the promotion of recycling.

The general outcome of this in-depth and detailed appraisal of the implication of plastic in the pollution of oceans and environments appears to be that the first and foremost goal to reach is to adhere to what the authors define as the "3R's" principles: reduce, redesign, and recycle. I would like to underscore the high scientific quality of this book, whose content is designed in a very didactic manner, clearly organized, and enriched by numerous illustrations, schemes, and diagrams. And that is supported by a solid bibliography following each one of the chapters.

I have no doubt that many categories of scientists, researchers, professionals, and students, as well as sociologists, policymakers, and innovators, will benefit from delving into this treatise. *Plastics in the Oceans: Toward Sustainable Solutions* should help pinpoint the most effective actions and create new solutions to foster sustainable practices.

Jean-Paul Joseleau, Emeritus Professor,
Grenoble Alpes University, Grenoble, France

Author biographies

Jean-Luc Wertz holds degrees in chemical engineering and economic science, as well as a PhD in applied science specializing in polymer chemistry, all from the Catholic University of Louvain, Belgium. He has held various international positions, serving as the Worldwide Director of R&D at Spontex and as the Technical Director of Edana. He holds several patents related to various industrial and consumer products. His expertise includes polymers, biobased products, biorefineries, and nonwovens. He has written seven books, five of them with CRC Press: *Hemicelluloses and Lignin in Biorefineries* (2018), *Starch in the Bioeconomy* (2020), *Chitin and Chitosans in the Bioeconomy* (2021), *Biomass in the Bioeconomy* (2022), and *Algae in the Bioeconomy* (2024).

Olivier Bédué received a degree in chemical engineering from the Ecole Nationale Supérieure Chimie de Paris (ENSCP) and a doctorate in polymer chemistry from the University Paris 6 (Jussieu), after which he worked for Spontex Company (Beauvais/France) as a R&D engineer and cellulose technical manager in industrial direction. He is now R&D Senior Manager (scourers/sponges) at the Regional Technical Center Europe for Freudenberg Home and Cleaning Solutions GMBH (Weinheim/ Germany). He is also a European Commission expert. He wrote two books in the field of the bioeconomy: *Cellulose Science and Technology* (EPFL Press, 2010) and *Lignocellulosic Biorefineries* (EPFL Press, 2013).

Serge Perez holds a doctorate in sciences from the University of Grenoble, France. He had international exposure throughout several academic and industry positions in research laboratories in the United States, Canada, and France (Centre de Recherches sur les Macromolecules Végétales, CNRS, Grenoble, as the chairperson and as Director of Research at the European Synchrotron Radiation Facility). His research interests span across the whole area of structural glycoscience, emphasizing polysaccharides, glycoconjugates, and protein-carbohydrate interactions. He has a strong interest in the economy of glycoscience and e-learning, for which he created www.glycopedia.eu; he is actively involved in scientific societies as president and past president of the European Carbohydrate Organisation. He is the author of more than 330 research publications, among which several are the subject of a large number of citations and references. He also edited and wrote several books, in particular: *Chitin and Chitosans in the Bioeconomy* in 2021, *The Biomass in the Bioeconomy* in 2022, and *Algae in the Bioeconomy* in 2024.

Acknowledgments

We express our deepest gratitude to Prof. Jean-Paul Joseleau, internationally recognized for his expertise in plant biochemistry and an emeritus professor at the University Grenoble Alpes, and Prof. Jacques Devaux, a highly recognized polymer scientist and engineer and emeritus professor at the Université catholique de Louvain, Belgium

Jean-Luc Wertz dedicates this book to his wife, Lydia; his two children, Vincent and Marie; and his four grandchildren, Mathilda, Nicolas, Carolina, and Laura.

Serge Perez dedicates this book to his grandchildren: Gabin, Sacha, Ana, Elya, and Roméo.

Finally, we thank Allison Shatkin, Ariel Finkel, and Deepika Ashok Kumar of the Taylor & Francis Group for their help and support. We also thank Jay Rajendiran for the production of this book.

1 Introduction

1.1 A BOOK TOWARD SUSTAINABLE PLASTICS

The second session of the Intergovernmental Negotiating Committee (INC) to develop an international legally binding instrument on plastic pollution, including in the marine environment, took place from 29 May to 2 June 2023 in Paris, France.[1] The session concluded with a mandate for the INC chair, with the support of the secretariat, to prepare a zero draft of the international agreement on plastic pollution ahead of the next session, due to take place in Nairobi, Kenya, in November 2023. In September 2023, the INC issued a zero draft document to be used as the basis of the global plastics treaty.

The growing pollution by plastics necessitates a structural shift.[2] Plastics are unique and indispensable because they are versatile and inexpensive materials with low weight and very good barrier properties. The demand for plastics, such as packaging, transportation, building, and construction, is constantly growing in many areas.

However, there are more and more problems associated with them: contribution to human CO_2 emissions, pollution in almost all environments, depletion of fossil resources, and planet sustainability. There is a planet versus plastics debate like the food versus fuel debate. We cannot produce more plastics forever or entirely stop their production. We must find a way toward sustainable, i.e., future-proof plastics. This can only be achieved by balancing optimal recycling/renewability and lower production and consumption.[3]

Global plastics production (polymers that are not used in converting plastic parts and products, i.e., used for textiles, adhesives, sealants, coatings, etc., are not included) was estimated at 390.7 million metric tons in 2021, an annual increase of 4%.[4] Plastic production has soared since the 1950s (Figure 1.1).[5]

It is projected that the global production of thermoplastics will amount to 445.25 million metric tons by 2025.[6] Annual production volumes are expected to continue rising in the following decades, rising to approximately 590 million metric tons by 2050.

Global cumulative production of plastics (referred to here as polymer resin and fibers) since 1950 is forecast to be 9.2 billion tons in 2017.[7] This figure is expected to increase to 37.5 billion tons in 2050.[8] Of the 9.2 billion tons, 5.3 billion tons are discarded, 1 billion tons are incinerated, and 2.9 billion tons (including 0.7 billion tons recycled) are still in use. Approximately 7 billion of the 9.2 billion tons of plastic produced from 1950 to 2017 became plastic waste, ending in landfills or dumping. Plastic production can grow faster due to the oil companies' strategies to replace missing fuel consumption with plastic production.

DOI: 10.1201/9781003532477-1

At least 14 million tons of plastic end up in the ocean yearly, making up 80% of all marine debris from surface waters to deep-sea sediments.[9] Marine species ingest or are entangled by plastic debris, which causes severe injuries and death. Plastic pollution threatens food safety and quality, human health, and coastal tourism and contributes to climate change. The plastic pollution crisis will not be solved by better recycling only; a systemic transformation toward a circular economy is needed.[10] Recycling alone will not be sufficient because most plastic is not designed today to be recycled and because there is a problem of contamination.[11]

According to the 5R's rule of waste management, four actions should be taken, if possible, before recycling: refuse, reduce, reuse, repurpose, and then recycle (Figure 1.2). It is also clear that we cannot continue using plastics as we have done and cannot live without them. We must develop a new, sustainable way of designing

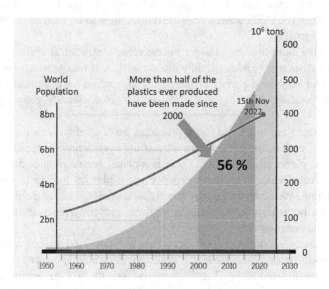

FIGURE 1.1 Global plastics production (adapted from Ref. Plastic Atlas [2019]).

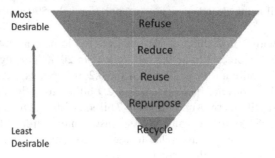

FIGURE 1.2 The 5R's of waste management.

and using plastics.[12] In the 19th century, Victor Hugo wrote: "How sad to think that nature speaks, and mankind does not listen."[13]

1.2 TYPES, VOLUMES, AND USES OF PLASTICS

Despite the large variety of polymers, eight of them make up 95% of all primary plastics ever made, which had exceeded 9 billion metric tonnes by the end of 2017: low-density polyethylene (LDPE); high-density polyethylene (HDPE); polypropylene (PP); polystyrene (PS); polyvinyl chloride (PVC); polyethylene terephthalate (PET); polyurethane (PUR); and polyester, polyamide, and acrylic (PP&A) (Figure 1.3 and Figure 1.4).[14]

With approximately 37%, packaging has the most significant share in global plastic production and makes up almost 50% of plastic waste due to its short service life (TNO, 2023). An Organisation for Economic Co-operation and Development (OCDE) report published in 2022 estimates that without radical actions, almost two-thirds of plastic waste in 2060 will be from short-lived items.[15] Looking for future-proof solutions, Stora Enso, a leader in the global bioeconomy, believes that the future of packaging will be circular, with all plastics being renewable and recyclable.[16]

The building and construction sector is the second biggest consumer of plastics, with around 16%. Moreover, the sector might cover more than 50% of the plastic stock in use due to their long service life.

Currently, the automotive sector produces 7% of global plastics, and the share of plastics in cars is expected to rise due to their light weight. Natural or synthetic rubber tires are one of the most important plastic applications in the transport sector.

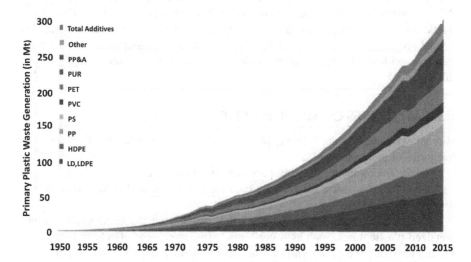

FIGURE 1.3 Global plastic waste production by type: LDPE, low-density polyethylene; HDPE, high-density polyethylene; PP, polypropylene; PS, polystyrene, PVC, polyvinyl chloride; PET, polyethylene terephthalate; PUR, polyurethane; PP&A, polyester, polyamide, and acrylic. (From Geyer et al., 2020; OECD, 2022.)

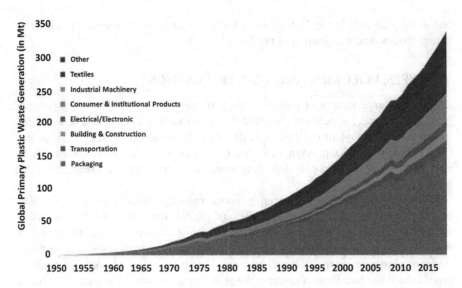

FIGURE 1.4 Global plastic waste production by application. (From Geyer et al., 2017; OECD, 2022.)

Despite their many benefits, agricultural plastics in agriculture pose a severe risk of pollution and harm to human and ecosystem health when damaged, degraded, or discarded in the environment. In 2019, agricultural value chains used 12.5 million tonnes of plastic products in plant and animal production.[17]

In textiles, plastics—namely synthetic polymer fibers—can offer significant benefits compared to natural fibers. However, textiles made of synthetic fibers cause plastic emissions during washing, drying, and use. Looking for future-proof textiles beyond recycling, UPM will produce new, climate-neutral materials from sustainably sourced forest biomass that will help replace fossil raw materials in the textile value chain.[18]

1.3 CATEGORIES OF MICROPLASTICS

Microplastics are fragments of any plastic less than 5 mm in length. There are two categories of microplastics:[19]

Primary microplastics are tiny particles designed for commercial use, such as microbeads in cosmetics, microfibers shed from clothing and fishing nets, debris from tire abrasion, etc.

Secondary microplastics are particles that result from the breakdown of more oversized plastic items, such as water bottles. In general, the effects of heat, light, air, and water are the most significant factors in the degradation of plastic polymers, leading to their oxidation and chain scission.

With the increasing use of plastics, more and more microplastics are formed. Primary and secondary microplastics are harmful to animals, humans, and the environment.

1.4 THE FATE OF PLASTIC IN THE OCEAN ENVIRONMENT

Understanding the fate of plastics in the environment is critically important for the quantitative assessment of the biological impacts of plastic waste.[20] In particular, there is a need to analyze the reputed longevity of plastics in the context of plastic degradation through oxidation and fragmentation reactions.

Since 1950, global plastic production and ocean plastic littering have increased exponentially.[21] Of the 359 million tonnes (Mt) produced in 2018, an estimated 14.5 Mt has entered the ocean. In particular, smaller plastic particles can be ingested by marine biota, causing hazardous effects. Plastic marine debris undergoes physical, chemical, and biological weathering, decreasing its size via fragmentation and altering its original shape, chemical composition, and surface characteristics (Figure 1.5). Most marine debris (80%) comes from land-based sources by wind, surface runoff, and rivers.[22]

1.4.1 PHOTO-OXIDATION AND FRAGMENTATION

Photo-oxidation (sometimes called oxidative photodegradation) is the degradation of a polymer surface due to the combined action of light and oxygen (Figure 1.6).[23] Photo-oxidation of plastic debris by solar UV radiation (UVR) makes material prone to subsequent fragmentation.

It is a chain process incorporating many chemical reactions after the outcome of the primary event—absorption of a photon, which induces breakdown to free-radical products.[24] It is the most significant factor in the weathering of plastics. Photo-oxidation results in the material becoming increasingly brittle and forming microplastics. Susceptibility to photo-oxidation varies depending on the chemical structure of the polymer. However, for many polymers, the photo-oxidation process can be

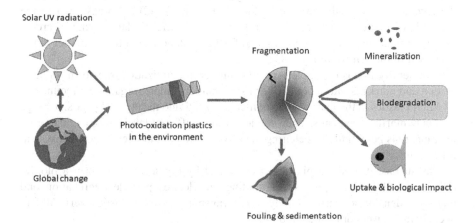

FIGURE 1.5 The fate of plastic in the ocean. Photo-oxidation of plastic by solar UV radiation makes it vulnerable to fragmentation and subsequent mineralization, biodegradation, uptake, fouling, and sedimentation. (From Andrady et al., 2022.)

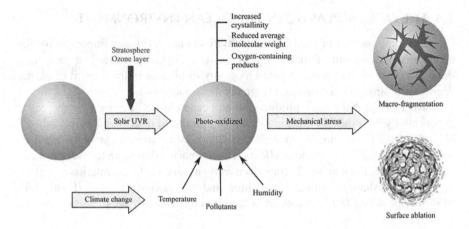

FIGURE 1.6 Schematic overview of photo-oxidative reactions, starting with a virgin plastic sphere (left) and leading to fragmentation (top right) and/or surface ablation (bottom right) of plastics. (From Andrady et al., 2022.)

divided into four stages: (1) initiation with the formation of free radicals on the polymer chain, which then react with oxygen; (2) propagation with the conversion of one active species to another; (3) chain branching resulting in side chains; and (4) termination in which active species are removed.[25]

1.4.2 Biodegradation, Mineralization, Uptake, Fouling, and Sedimentation

The biodegradation mechanism involves the action of microbial enzymes on the surface of the plastics.[26] Biodegradation typically comprises three distinct steps: (1) the formation of a biofilm on the surface of the plastic, (2) the breakdown of the plastic into smaller molecules through the action of extracellular enzymes secreted by microorganisms, and (3) ingestion and further metabolism of these smaller molecules within the cell (Audrady, 2022).

If mineralization is incomplete, organic matter mineralization transforms organic carbon and nutrients into inorganic forms or smaller, simpler organic compounds.[27]

Microplastics in the marine environment may fall into the optimal prey size range for many marine organisms. They can jeopardize different levels of biological organization, thus potentially threatening the conservation of biodiversity and ecosystem functioning.[28]

The surface of floating plastics is colonized by organisms that form a biofilm (Wayman, 2021). This process, biofouling, accelerates particle aggregation and increases density so particles may sink, transporting microplastics vertically to deeper water layers or the ocean floor.[29]

Plastic debris can also be colonized by living animals like barnacles and shrimps.[30]

1.5 PLASTIC WASTE TRADE

1.5.1 GLOBAL MARKET

The global plastic waste management market was estimated at USD 32.91 billion in 2019 and is expected to reach USD 41.58 billion by 2027, growing at a compound annual growth rate (CAGR) of 3.1% during the forecast period.[31] Based on the treatment method, the market is subdivided into collection, recycling, and disposal (Figure 1.7).

Asia-Pacific accounted for the largest revenue share in the market in 2019 and is projected to be the fastest growing at a CAGR of 4.0% during the 2019–2027 period. North America accounted for the second-largest revenue share in the market. Europe accounted for the third-largest revenue share in the market in 2019.

Key players are Waste Management, Inc. (USA), Suez (France), Veolia (France), Biffa Plc (UK), Clean Harbors, Inc. (USA), Covanta Holdings (USA), Hitachi Zosen (Japan), Remondis (Germany), Republic Services (USA), Stericycle (USA), ALBA Group (Germany), Recology (USA), TANA Oy (Finland), and Envac Group (Sweden).

1.5.2 PLASTIC WASTE IMPORTS AND EXPORTS

Around 2% of the world's plastic waste is traded nowadays. Most are traded within regions rather than between them (Figure 1.8).[32] However, unsustainable production and consumption of plastic have been supported by exporting waste to countries with lower energy and labor costs, with negative impacts on ecosystems, workers, and communities worldwide.[33] The global trade in plastic waste has mirrored the

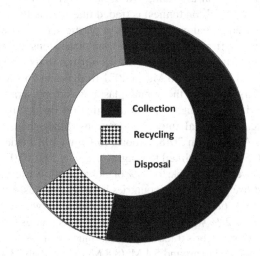

FIGURE 1.7 Global plastic waste management market share by treatment method. (Drawn with data from Fortune Business Insights, 2020.)

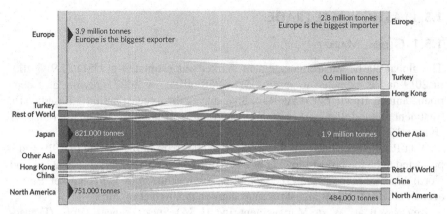

FIGURE 1.8 Plastic waste trade: Where does it come from, and where does it go? (Source: OECD [2022]. Monitoring trade in plastic waste and scrap. Based on United Nations Comtrade data. Licensed under CC-BY by the author Hannah Richie [Visual Capitalist, 2023].)

growth in global plastic production, allowing high-income, high-consuming countries to avoid the direct social and environmental impacts of their plastic problem and driving the ever-expanding production and consumption of virgin (new) plastics.

Every year, nations worldwide produce around 350 million metric tons (tonnes) of plastic waste. Most of this plastic waste is either incinerated or sent to landfills, thus eventually polluting our air, land, and oceans. Only a fraction of this waste is recycled, and just 2% (7 million tonnes) is traded internationally.

The annual reported global export weight of plastic scrap and waste fell by 50% over the past four years, from around 12.4 million metric tons (mega tonnes, Mt) per annum in 2017 to 6.2 Mt per annum in 2021. Compared to the previous year, global trade volume declined by 8.5% (0.5 Mt) in 2021 from around 6.7 Mt in 2020.[34]

Before 2018, a large share of the annual plastic waste and scrap exports was destined for China and Hong Kong. Exports to China declined substantially in 2018, triggered by tightened national import restrictions. Since then, exports to China decreased further. Although in 2018, exports to China still made up 7% of global exported weight (0.5 Mt), it declined to roughly 1.2% in 2021 (0.07 Mt) (Figure 1.9).

OECD countries were responsible for 89% of global plastic waste and scrap exports in 2021. Over the past five years, OECD countries have also increasingly become important importers: in 2021, 68% of all plastic waste and scrap was imported by OECD countries (in 2017, only 28% was imported). The OECD countries remain net exporters of plastic scrap and waste; debris trade surplus continues to decrease. Although in 2017, the trade surplus amounted to around 5.4 Mt (8.8 Mt exported and 3.4 Mt imported), in 2021, it was only 1.3 Mt (5.5 Mt exported and 4.2 Mt imported) (Figure 1.10).

With its reported plastic waste exports nearing 4 million metric tons, Europe is the largest exporting region in the world. However, as most is reportedly exported to other European nations, it is also the largest importing region (Visual Capitalist,

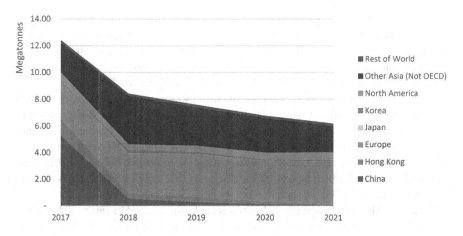

FIGURE 1.9 Annual export weight of waste, parings, and scrap of plastics by destination type, 2017–2021 (OCDE, 2023; Creative Common CC-BY).

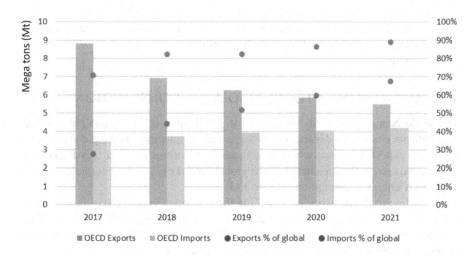

FIGURE 1.10 Annual import and export of waste, parings, and scrap of plastics reported by OECD member states (raw weight and share of global significance), 2017–2021 (OCDE, 2023; Creative Common CC-BY).

2023). According to UN Comtrade data, Germany, Japan, the United States, the UK, and the Netherlands are the world's top plastic waste exporters in 2020. In this order, Malaysia, Turkey, Germany, Vietnam, and the Netherlands have become the world's largest plastic waste importers.

1.5.3 REGULATIONS

The Basel Convention on the Control of Transboundary Movements of Hazardous Wastes and Their Disposal was adopted in 1989 and came into force in 1992.[35] It is the most comprehensive global environmental agreement on hazardous and other wastes.

As of November 2020, 187 countries and the European Commission are parties to the Convention. The Convention aims to protect human health and the environment against the adverse effects of generating transboundary movements and managing hazardous and other wastes. The Basel Convention forces parties to ensure that such wastes are collected and disposed of in an environmentally sound manner.

Aimed notably at addressing the gaps in plastic waste disposal, this treaty restricts participating nations from trading plastic scraps internationally unless they lack sufficient recycling or disposal capacity (Visual Capitalist, 2023). Furthermore, since 2015, the global plastic trade has indeed declined tremendously. However, millions of tonnes of plastics still need to be shipped; most are mismanaged. In May 2019, amendments were adopted, which mean that, as of January 2021, plastic waste that is difficult to recycle will need to be consented to before being imported into receiving countries (EIA, 2021). Although a move in the right direction, these amendments will still allow Global North countries to dump difficult-to-recycle plastic waste in the Global South, where there are often no infrastructure and capacity to manage it appropriately.

In November 2021, the European Commission proposed a new regulation on waste shipments.[36] It aims to ensure that the EU does not export its waste challenges to third countries and supports a clean and circular economy. The proposal plans (1) to establish new rules for EU waste exports, (2) to make it easier to transport waste for recycling or reuse in the EU, and (3) to set out new measures to tackle illegal waste shipments.

1.6 NEED TO REDUCE FOSSIL RESOURCES CONSUMPTION

The world economy is highly dependent on fossil fuels, oil, coal, and natural gas. A significant cost of this dependence is that energy-producing countries can use their fossil fuel exports to pressure or threaten energy importers. Another massive cost of this dependence is climate change at the global level. Fossil fuels generate over 80% of primary energy consumption (Figure 1.11).[37]

According to the Network for Greening the Financial System (NGFS), this share must be reduced to around 30% to reach net-zero emissions by 2050. For the EU, the reduction will have to be even greater. It will require wide-ranging structural changes in energy production and economic systems.

Plastics are made from fossil fuels, accounting for 4–8% of global oil consumption.[38] Net-zero emission plastics are possible and technologically feasible. A team of scientists from Germany, Switzerland, and the United States has shown how the raw material combination of biomass, CO_2, and recycled waste can be used to produce plastic with net-zero emissions. Renewable raw materials plus renewable energy are keys toward future-proof plastics. Cleaning the current pollution is the other challenge.

1.7 STRATEGIES TO PRODUCE FOSSIL PLASTICS
AS AN ALTERNATIVE TO FOSSIL FUELS

Plastics show the most substantial production growth of all bulk materials and are responsible for 4.5% of global greenhouse gas (GHG) emissions.[39] The carbon

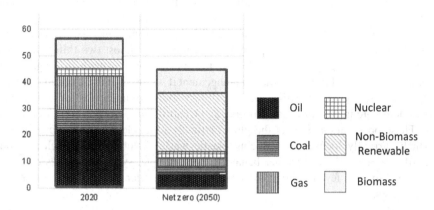

FIGURE 1.11 Global energy mix, exajoules (10^{18}J) per year. Net zero by 2050 is an ambitious scenario that limits global warming to 1.5 °C, reaching net-zero emissions by 2050. (Redrawn with data from "Network for Greening the Financial System (NGFS)." European Central Bank, 2022.)

footprint of plastics reached 2 GtCO$_2$-equivalent (CO$_2$e) in 2015, accounting for 4.5% of global GHG emissions.

Without implementing new policies, experts project a doubling of global plastic demand by 2050 with an almost equivalent increase in CO$_2$ emissions. Since their recent record profits, oil and gas companies have invested massively in the production of plastics.

In June 2023, Saudi Arabian Oil Company (Aramco) and TotalEnergies signed an $11 billion contract to start building a new petrochemicals complex in Saudi Arabia.[40] The award of engineering, procurement, and construction contracts for main process units and associated utilities marks the start of construction work on this joint project.[41] Integrated with the SATORP existing refinery in Jubail, the new petrochemical complex will house a giant mixed-load steam cracker in the gulf, with a capacity to produce 1.65 million tons of ethylene, a major raw material for plastics and other industrial gases per year.

In a world where the fight against climate change has become a global priority and thermal cars will disappear, oil and gas companies need to find new outlets. According to international experts, the strategy of many oil companies appears to produce fossil plastics as an alternative to fossil fuels, condemned to lower consumption in the long term.[42] Therefore, plastic production is planned to increase by about 30% from 2025 to 2050, and global plastic waste is set to almost triple by 2060.[43]

1.8 CLOTHES AND TIRES AS SIGNIFICANT SOURCES OF MICROPLASTIC POLLUTION

Primary microplastics washed off products such as synthetic clothes and car tires could contribute up to 30% of the "plastic soup" polluting the world's oceans and—in many developed countries—are a more significant source of marine plastic pollution than plastic waste.[44] Sources of primary microplastics include car tires, synthetic textiles, marine coatings, road markings, personal care products, plastic pellets, and city dust.

Between 15% and 31% of the estimated 9.5 million tonnes of plastic released into the oceans each year could be primary microplastics, almost two-thirds of which come from washing synthetic textiles and the abrasion of tires while driving. Therefore, we must look far beyond waste management if we are to address ocean pollution in its entirety. Synthetic textiles are the primary source of microplastics in Asia, and tires dominate in the Americas, Europe, and Central Asia.

These observations impact the global strategy to tackle ocean plastic pollution, which currently focuses on reducing plastic waste. Solutions must include product design as well as consumer behavior. For example, synthetic clothes could be designed to shed fewer fibers, and consumers could prefer natural fabrics over synthetic ones.

1.9 TOXICITY FOR ECOSYSTEMS, ANIMALS, AND HUMANS

Microplastic pollution has recently been identified as a significant issue for the health of ecosystems such as mangroves and coral reefs.[45] Coral reefs are losing the capacity to sustain their fundamental ecosystem functions, with plastic pollution adding to the threats from climate change and overfishing.[46] The mangrove ecosystem, a buffer between the land and the sea, has been identified as a potential sink of microplastics.[47] Wastewater treatment plants, aquaculture, and in situ degradation of plastics are the primary sources of microplastic pollution in mangroves.

Plastic harms animals.[48] They eat it, they get caught in it, or they get sick because of it. Substantial impacts of plastic debris are the ingestion, suffocation, and entanglement of hundreds of marine species.[49] Marine wildlife such as seabirds, whales, fish, and turtles mistake plastic waste for prey; most then die of starvation as their stomachs become filled with plastic. They also suffer from lacerations, infections, reduced ability to swim, and internal injuries. Floating plastics also help transport invasive marine species, threatening biodiversity and the food web.

Human exposure to microplastics has been confirmed by their presence in multiple tissues.[50] Microplastics can affect the human body by stimulating the release of endocrine disruptors, which can harm the body by causing various cancers and reproductive system disorders.[51] In addition, microplastics can carry other toxic chemicals, such as heavy metals and organic pollutants, during adsorption, which can adversely affect the final consumer (Figure 1.12).

1.10 STRUCTURE OF THE BOOK

Chapter 1 introduces the book and includes its purpose and the present status of plastics with their types, amounts and uses, the microplastics, their degradation in the ocean, their trade, the need to reduce fossil consumption, the new strategies towards more plastics, significant sources of pollution and their toxicity.

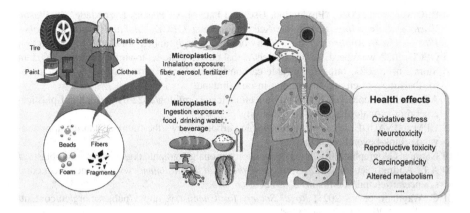

FIGURE 1.12 Health effects of microplastic exposure (Lee, 2023, Yonsei Medical Journal; CC-BY).

Chapter 2 deals with the problem of marine plastic pollution, including sources, plastic pathways, distribution in the ocean, ecological and biological impacts, implications for well-being, causes, and drivers.

Chapter 3 presents the future of plastic pollution, including new policies, public awareness, reduced consumption versus recycling, alternative materials, upstream preventive solutions, possible scenarios, and a vision for sustainable plastics.

Chapter 4 concludes the book with strategies toward future-proof plastics.

NOTES

1 UN Environment Programme, 2023, Paris, www.unep.org/events/conference/second-session-intergovernmental-negotiating-committee-develop-international/media#PressRelease
2 Renewable Carbon News, 2023, https://renewable-carbon.eu/news/white-paper-new-guide-to-the-future-of-plastics/
3 Wevolver, 2022, www.wevolver.com/article/future-proofing-plastics
4 Statista, June 12, 2023, www.statista.com/statistics/282732/global-production-of-plastics-since-1950/
5 Plastic Atlas, 2019, 2nd Edition.pdf, boell.de
6 Statista, March 24, 2023, www.statista.com/statistics/664906/plastics-production-volume-forecast-worldwide/
7 UN Environment Programme, 2023, www.unep.org/plastic-pollution
8 Statista, 2023, file:///C:/Users/User/Documents/Plastic%20cumulative%20production%20globally%202050%20Statista.htm
9 IUCN, International Union for Conservation of Nature, 2021, www.iucn.org/resources/issues-brief/marine-plastic-pollution
10 www.unep.org/es/node/31139
11 https://just-zero.org/our-stories/blog/plastic-pollution-problem-cant-recycle-our-way-out/
12 TNO, 2023, www.tno.nl/en/newsroom/2023/06/from-plastic-free-future-proof-plastics/
13 Victor Hugo, www.buboquote.com/en/quote/3503-hugo-how-sad-to-think-that-nature-speaks-and-mankind-doesn-t-listen

14 R. Geyer, et al., 2020, "Production, Use, and Fate of All Plastics Ever Made", in *Plastic Waste and Recycling*, pp. 13–32, Academic Press; OECD, *Global Plastics Outlook—Plastics Use by Application*, https://ourworldindata.org/grapher/global-plastics-production

15 OECD, 2022, www.oecd.org/environment/global-plastic-waste-set-to-almost-triple-by-2060.htm

16 Stora Enso, 2023, https://renewable-carbon.eu/news/stora-enso-is-laying-the-foundation-for-the-future-of-circular-packaging-in-oulu-finland/

17 Alliance for Science, 2022, https://allianceforscience.org/blog/2022/01/agricultural-plastics-emerging-as-major-threat-to-sustainability/

18 UPM and VAUDE, 2023, https://renewable-carbon.eu/news/the-future-of-fabrics-upm-and-vaude-think-beyond-recycling/

19 National Geographic, 2022, https://education.nationalgeographic.org/resource/microplastics/

20 A.L. Andrady, et al., 2022, *Science of the Total Environment*, www.sciencedirect.com/science/article/pii/S004896972205121X

21 C. Wayman, et al., 2021, *Royal Society for Chemistry*, https://pubs.rsc.org/en/content/articlehtml/2021/em/d0em00446d

22 Clean Water Action, 2023, https://cleanwater.org/problem-marine-plastic-pollution

23 https://en.wikipedia.org/wiki/Photo-oxidation_of_polymers

24 www.sciencedirect.com/topics/chemistry/photooxidation

25 www.sciencedirect.com/topics/engineering/photooxidation

26 M. Srikanth, 2022, *Bioresources and Bioprocessing*, https://bioresourcesbioprocessing.springeropen.com/articles/10.1186/s40643-022-00532-4

27 S.D. Bridgham, 2013, https://acsess.onlinelibrary.wiley.com/doi/abs/10.2136/sssabookser10.c20

28 C. Corinaldesi, 2021, *Nature*, www.nature.com/articles/s42003-021-01961-1

29 L.A. Amaral-Zettler, 2021, *Water Research*, www.sciencedirect.com/science/article/pii/S0043135421004875

30 The Guardian, 2021, *Coastal Species are Forming Colonies on Plastic Trash*, www.theguardian.com/us-news/2021/dec/09/coastal-species-are-forming-colonies-on-plastic-trash-in-the-ocean-study-finds

31 FORTUNE Business Insights, 2020, www.fortunebusinessinsights.com/plastic-waste-management-market-103063

32 Visual Capitalist, 2023, www.visualcapitalist.com/cp/charting-the-movement-of-global-plastic-waste/

33 Environmental Investigation Agency, 2021, https://reports.eia-international.org/a-new-global-treaty/plastic-waste-trade/

34 A. Brown, et al., 2023, *Monitoring Trade in Plastic Waste and Scrap*, OECD, www.oecd.org/environment/monitoring-trade-in-plastic-waste-and-scrap-39058031-en.htm

35 UN Environment Programme, *Basel Convention on the Control of Transboundary Movements of Hazardous Wastes*, www.unep.org/resources/report/basel-convention-control-transboundary-movements-hazardous-wastes

36 European Commission, https://environment.ec.europa.eu/topics/waste-and-recycling/waste-shipments_en

37 F. Panetta, 2022, *European Central Bank*, www.ecb.europa.eu/press/key/date/2022/html/ecb.sp221116~c1d5160785.en.html

38 M. Steilemann, 2022, *The Davos Agenda*, www.weforum.org/agenda/2022/01/it-s-time-to-shift-to-net-zero-emissions-plastics/

39 P. Stegmann, 2022, *Nature*, www.nature.com/articles/s41586-022-05422-5

40 Reuters, 2023, www.reuters.com/business/energy/aramco-totalenergies-sign-11-bln-contract-build-petrochemicals-complex-saudi-2023-06-24/

41 TotalEnergies, 2023, https://totalenergies.com/media/news/press-releases/aramco-and-totalenergies-award-contracts-11-billion-amiral-project

42 A. Dumont, 2023, *Zero Waste France*, www.zerowastefrance.org/face-cachee-industrie-plastique/
43 OECD, 2022, *Global Plastic Waste to Almost Triple by 2060*, www.oecd.org/environment/global-plastic-waste-set-to-almost-triple-by-2060.htm
44 International Union for Conservation of Nature and Natural Resources, 2017, www.iucn.org/news/secretariat/201702/invisible-plastic-particles-textiles-and-tyres-major-source-ocean-pollution-%E2%80%93-iucn-study
45 J. John, 2022, *Environmental Chemistry Letters*, Springer, https://link.springer.com/article/10.1007/s10311-021-01326-4
46 P. Stefanoudis, 2023, University of Oxford, www.ox.ac.uk/news/2023-07-13-new-study-finds-plastic-pollution-be-almost-ubiquitous-across-coral-reefs-mostly
47 H. Deng, 2021, *Science of the Total Environment*, www.sciencedirect.com/science/article/abs/pii/S0048969720355704
48 Plastic Soup Foundation, www.plasticsoupfoundation.org/en/plastic-èproblem/plastic-affect-animals/
49 International Union for Conservation of Nature and Natural Resources, 2021, www.iucn.org/resources/issues-brief/marine-plastic-pollution
50 J.C. Prata, 2023, *Animals*, www.ncbi.nlm.nih.gov/pmc/articles/PMC9951732/
51 Y. Lee, 2023, *Yonsei Medical Journal*, www.ncbi.nlm.nih.gov/pmc/articles/PMC10151227/

2 The Problem of Marine Plastic Pollution

2.1 SOURCES AND EXTENT OF MARINE PLASTIC POLLUTION

Plastic pollution in our marine environment occurs with more than 9.5 million tons of new plastic waste flowing into the ocean yearly (more than 8 million tons from macroplastics capable of generating secondary microplastics and more than 1.5 million from primary microplastics).[1] This is impacting our planet's precious biodiversity and damaging the ecosystems. The widespread contamination of our oceans is also fast becoming a worldwide human health risk as plastic enters our food and water supplies. This section provides a global estimate and mapping of the sources and quantities of primary microplastics (plastics that enter the oceans in the form of tiny particles released from household and industrial products). Although mismanaged plastic waste is still the primary source of marine plastic pollution globally, it shows that, sometimes, more plastic may be released from our driving and washing activities than from the mismanagement of our waste. How we currently design, produce, consume, and dispose of plastic appears unsustainable and inefficient.[2] Solutions should come from innovative materials, smart design, and, importantly, public awareness to change consumption and disposal habits, with these efforts being sustained by binding regulations. The challenge is future-proofing plastics on a healthy planet.

2.1.1 PRIMARY MICROPLASTICS

Primary microplastics are globally responsible for a major source of plastics in the oceans. Between 15% and 31% of ocean plastic could originate from primary sources (International Union for Conservation of Nature, 2017). Where advanced waste treatment facilities exist, primary microplastic releases even outweigh secondary microplastics from the degradation of large plastic waste (Figure 2.1).

The global release of primary microplastics into the ocean was estimated at 1.5 million tons per year (Mtons/year), ranging between 0.8 and 2.5 Mtons/year. These releases are a fraction of the estimated losses of primary microplastics into the environment, varying between 1.8 and 5.0 Mtons/year (central: 3.2 Mtons/year). Most of these primary microplastics stem from laundering synthetic textiles and tire abrasion while driving. Other sources include city dust, road markings, marine coatings, personal care products, and plastic pellets.

2.1.1.1 Synthetic Textiles: Abrasion during Laundry

Washing synthetic textiles in industrial and consumer laundries creates primary microplastics via abrasion and shedding of fibers. Fibers are then discharged into sewage water and potentially into the ocean using a wastewater pathway. There is increasing evidence available that microplastic fibers released during washing were

DOI: 10.1201/9781003532477-2

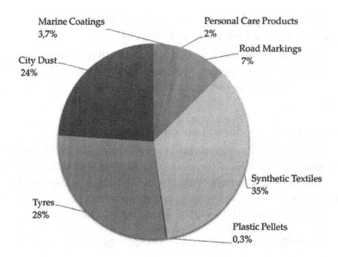

FIGURE 2.1 Global release of primary plastics to the world oceans: 1.5 million tons per year (K. Schhen et al., 2019, Plastic and Environment. Creative Commons Attribution 3.0 Unported License).

likely formed during the manufacturing stage.[3] Primary microplastics from synthetic textiles are estimated to be 35% of global releases to the ocean (IUCN, 2017).

2.1.1.2 Tires: Abrasion while Driving

Tires get abrased during use. The particles are formed from the outer parts of the tire and consist of a matrix of polymers such as styrene-butadiene rubber (SBR) with many other additives. Tire dust will then be spread by the wind or washed off the road by rain and potentially released into the ocean. Primary microplastics from tires are estimated to be 28% of global releases to the sea (IUCN, 2017).

2.1.1.3 City Dust: Weathering, Abrasion, and Pouring

City dust refers to a wide range of microplastic sources originating from urban areas.[4] Weathering, abrasion, and detergents create city dust from manufactured products. City dust includes losses from the abrasion of objects (synthetic soles of footwear, synthetic cooking utensils), the abrasion of infrastructure (household dust, city dust, artificial turfs, harbors and marinas, building coating), and the blasting of abrasives and intentional pouring (detergents). Primary microplastics from city dust are estimated to be 24% of global releases to the ocean (IUCN, 2017).

2.1.1.4 Road Markings: Weathering and Abrasion by Vehicles

Different types of markings (paint, thermoplastic, preformed polymer tape, and epoxy) are applied, with a dominance of paint. Loss of microplastics may result from weathering or abrasion by vehicles. As for tires, dust will either be spread by the wind or washed off the roads by rain before reaching permanent surface waters and potentially the oceans. Primary microplastics from road markings are estimated to be 7% of global releases to the sea (IUCN, 2017).

2.1.1.5 Marine Coatings: Weathering and Incidents during
Application, Maintenance, and Disposal

Marine coatings are applied to all parts of vessels for protection. They include solid coatings, anticorrosive paint, and antifouling paint. Primary microplastics from marine coatings are estimated to be 3.7% of global releases to the ocean (IUCN, 2017).

2.1.1.6 Personal Care Products: Loss during Use

Plastic microbeads are used as ingredients in personal care for various purposes. These ingredients represent up to 10% of the product weight and several thousand microbeads per gram. A classical use of personal care products results in the introduction of plastic particles into wastewater streams. Primary microplastics from personal care products are estimated to be 2% of global releases to the ocean (IUCN, 2017).

2.1.1.7 Plastic Pellets: Incidents during Manufacturing,
Transport and Recycling

Plastic pellets are the raw materials for plastic transformers that generate the plastic products. During manufacturing, processing, transportation, and recycling, pellets can be spilt into the environment through incidents. Primary microplastics from plastic pellets are estimated to be 0.3% of global releases into the ocean (IUCN, 2017)/

2.1.2 PLASTIC WASTE SOURCE OF SECONDARY MICROPLASTICS

Ocean plastic waste includes food wrappers, beverage bottles, grocery bags (single-use bags), straws, and take-out containers.[5]

2.1.2.1 Statistics and Facts

A few astonishing statistics regarding the plastic waste problem in the ocean are given next[6]:

- Fifty percent of all the plastic in the world's oceans is single-use plastic waste; only 9% is recycled plastic.
- A total of 500 billion (10^9) plastic bags and slightly fewer plastic bottles are used yearly worldwide.
- Every year, 79% of plastic ends up in landfills or the oceans, 9% is recycled, and 12% is industrially incinerated.
- The *Great Pacific Garbage Patch* between Hawaii and California measures around 1.6 million square kilometers, triple the size of France.
- More than $1 billion worth of plastic is wasted each year.
- There will be more plastic than fish in the ocean in 2050, according to the Ellen MacArthur Foundation.
- More than 1 million seabirds and 100,000 marine animals yearly die from plastic pollution.
- One hundred percent of baby sea turtles have plastic in their stomachs.
- There are now 5.25 trillion (10^{18}) macro and micro pieces of plastic in our ocean, weighing up to 269,000 tonnes.

- There is an estimated 75–199 million tonnes of plastic waste currently in our oceans, with a further 8–14 million tonnes of plastic entering the marine environment every year.
- Eighty-eight percent of the sea's surface is polluted by plastic waste.
- Plastic generally takes between 500 and 1000 years to degrade. Even then, it becomes microplastics without fully degrading.
- Eighty percent of ocean trash consists of plastics, while three-quarters of ocean trash comprise just ten items, the top eight of which are plastic.
- The top four items that pollute the ocean are plastic bags, making up 14% of the trash; plastic bottles, comprising 12%; plastic food containers and cutlery, which account for 9%; and wrappers, which also make up 9%.

2.1.2.2 Central Issue of Recycling

According to Neste, a leader in renewable solutions, some 353 million tons of plastic waste were generated worldwide in 2019.[7] It is estimated that by 2060, the figure will increase to a billion tons. Just 9% of global plastic waste is being recycled (Figure 2.2).

About 14% of plastic waste is collected for recycling. About half of the waste collected for recycling ends up in landfills, incinerated, or mismanaged. Furthermore, plastic waste differs in composition (e.g., single-material vs. multiple materials) and color. All these differences lead to some plastic waste being relatively easy to recycle while other debris is not. Hence, a sorting facility using different processes is necessary. Such a facility will be followed for easy-to-recycle plastics by an extruder to generate recycled pellets. Plastics with multiple materials will instead go to a chemical recycling facility, where they will be disassembled into their building blocks.

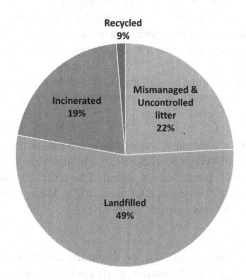

FIGURE 2.2 Only 9% of plastic waste is today recycled globally from more than 350 million tons of plastic waste generated in total per year (Neste, 2023. Redrawn from OECD data [2022]).

2.1.2.3 Plastic Bags

Plastic bags start as fossil fuels and become waste in landfills and the ocean.[8] Birds often mistake shredded plastic bags for food. Sea turtles frequently cannot distinguish between jellyfish and floating plastic bags. Once eaten, these soft plastics block their stomachs. Fish eat thousands of tons of plastic annually, transferring it up the food chain. The fossil fuel industry plans to increase plastic production by 40% over the next decade. These oil and gas giants are rapidly building petrochemical plants to turn fossil fuels into plastics. This leads to more plastics in the ocean, more greenhouse gas emissions, and worse climate change.

Plastic bags are the most common plastic waste in the oceans, accounting for 14% of all ocean trash (Vetter, Forbes, 2023; Morales Caselles, 2021). Most plastic bags are made from high-density polyethylene (#2 plastic), but thinner-material bags (such as produce bags) are made from low-density polyethylene (#4 plastic). The recycling collection system is widely available, primarily through collection bins. Recycling a ton of plastic bags saves 11 barrels of oil.

The EU has a Plastic Bags Directive (Directive 2015/720), an amendment to the Packaging and Packaging Waste Directive (Directive 94/62/EC). It was adopted to deal with the unsustainable consumption of lightweight plastic carrier bags.[9]

2.1.2.4 Plastic Bottles

Plastic bottles are single-use plastic, designed to be used only once and discarded.[10] After their use, the bottles can be left:

- In the recycling bin: Bottles destined for recycling are unlikely to end up in the ocean in their current form unless they are mismanaged or lost in transit to a processing facility. Due to recent international limitations, many of these bottles will unfortunately end up in landfills.
- In the trash can: These bottles also will not likely end up, in their current form, in the ocean. However, with poor waste management, some particles from the bottles may seep into the soil, eventually entering waterways and, ultimately, the ocean.
- As litter: Bottles dropped as litter may very well be carried by wind, rain, or other processes to sewers, rivers, and other waterways, ultimately bringing the bottle into the ocean.

Plastic bottles are the second most common plastic waste in the oceans, making up 12% of all ocean trash (Vetter, Forbes, 2023; Morales Caselles, 2021). Less than half of the plastic bottles purchased are collected for recycling, and just 7% of those collected are turned into new bottles (Optimum, 2023). Most plastic bottles end up in landfills or the ocean without coherent alternatives.

2.1.2.5 Fishing Gear

Known officially as abandoned, lost, or discarded fishing gear (ALDFG)—and unofficially as "ghost gear"—this marine waste comprises fishing nets, ropes, lines, traps, and other fishing paraphernalia, mostly made of durable plastics.[11] Highly buoyant plastic fishing gear is more likely to become concentrated in places such as the North Pacific gyre, but it is also dispersed across the ocean. The quantity is notoriously difficult to measure, but between 500,000 and 1 million tons a year end up in the oceans. It has been estimated that ALDFG comprises a significant part (at least 46%) of the 79,000 tons of plastic waste accumulated in the Great Pacific Garbage Patch.[12]

Experts say that ALDFG is the deadliest form of marine plastic, threatening 66% of marine animals, including all sea turtle species, and 50% of seabirds.[13] ALDFG does not decompose quickly in the sea, but eventually breaks into small pieces and becomes secondary microplastics. In the case of a fishing line made of nylon, for example, the time for natural decomposition would be over 600 years.

2.1.2.6 Textile Wastes

Big brands are dumping millions of tonnes of used clothing and textiles of such poor quality that they are immediately sent to dump sites—polluting the environment, having social impacts, posing health risks, and contributing to the climate crisis.[14]

Africa is the center of this growing textile waste crisis, and Ghana is the hub for clothing used abroad in West Africa.[15]

The fashion industry is one of the biggest polluters of freshwater sources due to used clothing that enters river systems, often ending up in oceans, from where it can get washed back up on beaches, endangering marine life and ecosystems.

The disparity between the Global North's and Global South's sustainability efforts is a significant issue. Fast fashion brands, in particular, have been criticized for prioritizing environmental responsibility in their home markets while disregarding the consequences in developing nations.

The Global North is increasingly adopting sustainable practices, such as using recycled materials and reducing water usage. However, sustainability is often an afterthought in the Global South, where these brands source materials and produce clothing.

2.1.2.7 Plastic Mulch

Plastic mulching is the process of covering the earth around a plant with a plastic film to help it grow. It is used in agriculture, farming, and gardening. There are two basic types of plastic mulching: black polyethylene (PE) film and clear PE.[16] The global annual use of plastic mulching is 4 million tons, increasing 5.6% annually.[17]

There are no consistently accessible mechanisms for sustainably disposing of plastic mulch film at the end of its life.[18] Furthermore, pieces of this plastic can break down into the soil and waterways as secondary microplastics, which present severe health and ecosystem concerns and may ultimately reach the ocean.

PE mulch has been used in agriculture since the 1950s to improve weed management, reduce soil water loss, raise soil temperature, increase yield, enhance crop quality, and shorten harvest time.[19] Soil-biodegradable mulch has been developed as an environmentally friendly alternative to PE mulch. The most common biobased feedstocks used to make biodegradable plastic mulches are starch, polylactic acid (PLA), and polyhydroxyalkanoates (PHA).

2.2 LIFE CYCLE OF PLASTICS AND PATHWAYS FROM LAND TO OCEAN

2.2.1 LIFE CYCLE OF PLASTICS

Primary microplastic loss occurs at various stages of the plastic lifecycle (Figure 2.3) (IUCN, 2017). Plastic waste occurs during the product disposal stage, which can be managed (incineration and landfilling) or mismanaged (including uncontrolled littering).

Plastic pellets are the only losses occurring during plastic production, transport, or recycling stages. Most losses happen during the use phase of products. On the contrary, secondary microplastics mainly originate from the breakdown of mismanaged waste during the disposal stage.

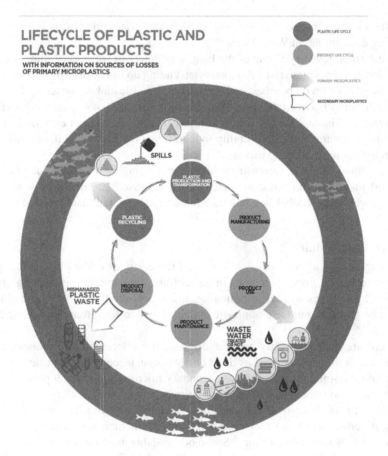

FIGURE 2.3 Life cycle of plastic products showing the losses of primary microplastics and the production of plastic waste (International Union for Conservation of Nature, 2017).

2.2.2 MICROPLASTICS PATHWAYS FROM LAND TO OCEAN

Most (98%) of primary microplastic losses stem from land-based activities (Figure 2.4).

The main pathway from land to ocean and soil is road runoff (66% of land-based losses), followed by wastewater treatment systems (25%) and wind transfer (7%). Marine activities only generate around 2% of the losses. All losses occurring in the ocean and all losses transported by wind become releases. Regarding ocean releases, 44% are along the road runoff pathway, 37% are along the wastewater pathway, 15% are transported by wind, and 4% are direct releases.

2.2.2.1 Road Runoff Pathway

Some significant sources of plastics from road runoff include tire abrasion, road markings, and plastic pellets.[20] Rural areas are a minor contributor to ocean plastic waste in developed countries. At the same time, urban runoff is often collected in sewer systems and can either be dumped directly into aquatic environments or enter wastewater pathways. Road runoff is the most extensive microplastic distribution pathway, accounting for 66% of the total, and of this, an estimated 32% ends up as ocean releases and an estimated 68% ends up as soil releases.

Potential mitigation methods for road runoffs include permanent stormwater retention ponds, sediment traps, diversion trenches, grass strips, and slope protection. However, this will require substantial, costly infrastructure development worldwide.

2.2.2.2 Wastewater Pathway

Microfibers are released from fabrics during the laundering process. They form a significant part of primary microplastics in the ocean. The most common forms of microplastics that enter wastewater pathways include personal care products and cosmetics,

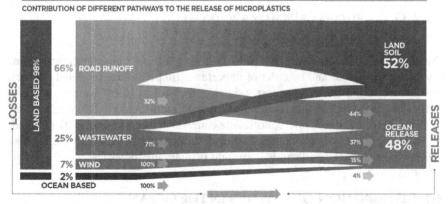

FIGURE 2.4 Global releases to the world oceans. Contribution of different pathways to the release of microplastic waste (International Union for Conservation of Nature, 2017).

abrasion of plastic products and tires, textiles, and road marking paints (Watt, NLM, 2021). Challenges with microfiber leakage exist because municipal water is discharged into the environment without any treatment procedures in many areas. Estimates made by the United Nations suggest that as much as 80% of global wastewater remains untreated once released back into the environment. Implementing wastewater treatment plants has been gaining interest in combating the accumulating amounts of microplastics. It could be a viable solution to reduce the gross microplastic increase in aquatic environments. Another technique to diminish microplastics is using effective washing machine filters, which consumers can quickly adopt to reduce microplastic fibers.

2.2.2.3 Wind Transfer
The primary sources of wind pathways are tires and road markings. The wind is estimated to transfer 7% of the land-based losses to the ocean.

2.2.3 Plastic Waste Mismanagement

Over 80% of the plastic debris found in marine environments is traced back to land-based sources, with the most significant contributor stemming from industrialized areas due to solid waste disposal, littering, and plastic usage (Watt, 2021). This arises from the low residual value and the small size of many plastic packaging products, resulting in over 8 Mt of plastic leakage into the ocean annually. Improper waste management is most prevalent in middle-income countries (Figure 2.5). The figure displays a projected worldwide map for global mismanaged waste for 2025 based on total predictions of 69.14 Mt of mismanaged plastic waste.

The figure highlights the impact of adequate waste management infrastructure, as waste produced within 50 km of coastlines is considered at high risk of entering oceans. Plastics generated in populations beyond 50 km of a coastline are deemed unlikely to travel into ocean environments, and of the identified mismanaged waste, 15–40% is estimated to become marine debris.

2.2.4 Ocean Releases of Primary Microplastics
and Plastic Waste by Regions
Critical sources of ocean releases of primary microplastics differ among regions. The key issues, by regions and by order of importance (in percentage of global releases) are, textiles in India and Southeast Asia (15.9%), tires in North America (11.5%), textiles in China (10.3%), and tires in Europe and Central Asia (10.3%) (Figure 2.6).

Releases from primary microplastics are equivalent to or outweigh that of secondary microplastics from mismanaged waste in Europe and North America. In other regions, plastic waste significantly outweighs primary microplastics.

2.3 DISTRIBUTION OF PLASTICS IN THE OCEAN

2.3.1 Ocean Garbage Patches

A portion of the plastic that enters the ocean travels to ocean garbage patches.[21]

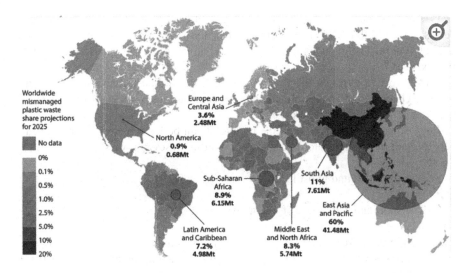

FIGURE 2.5 Projections for worldwide mismanaged plastic waste generated within 50 km of coastlines in 2025, with share contributions by each country indicated based on a total projected mismanaged waste of 69.14 Mt. Estimates predict anywhere between 15% and 40% of these mismanaged wastes will be converted to marine debris (Watt, 2021).

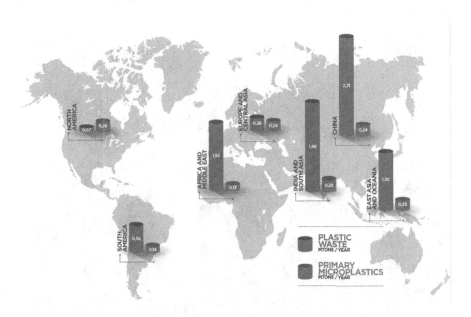

FIGURE 2.6 Ocean releases, by regions, of primary microplastics and plastic waste in Mtons/year (International Union for Conservation of Nature, 2017).

Garbage patches are large ocean areas where litter, fishing gear, and other marine debris collect.[22] Rotating ocean currents then form "gyres," giant whirlpools that suck objects in. The gyres pull trash into one location, often the gyre's center, creating "patches."

There are five gyres in the ocean. One is in the Indian Ocean, two are in the Atlantic Ocean, and two are in the Pacific Ocean. Garbage patches of varying sizes are located in each gyre (Figure 2.7).

Because garbage patches are quite remote, it can be hard to study them. Marine debris found in garbage patches can impact wildlife in several ways:

- Entanglement and ghost fishing
- Ingestion
- Transport and settlement of non-native species

Cleaning the ocean from garbage patches is a highly challenging task. Large debris, like fishing nets, can be removed by people, but mixing and spreading microplastics may be very difficult.

2.3.2 GREAT PACIFIC GARBAGE PATCH

The most famous of these patches is often called the "Great Pacific Garbage Patch" (GPGP), also known as the Pacific Trash Vortex. It was first discovered by Captain

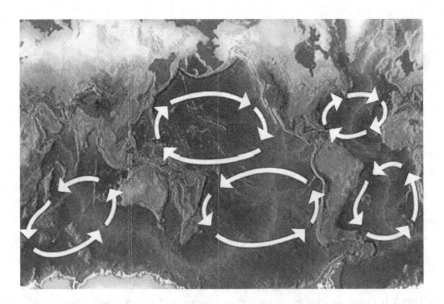

FIGURE 2.7 The five ocean gyres (from Moreau René, 24 July 2020). The slow and powerful ocean circulation. (From Encyclopedia of the Environment, 2020, www.encyclopedie-environnement.org/en/water/slow-powerful-ocean-circulation. Creative Commons BY-NC-SA license.)

Charles Moore in 1997 when sailing back after a yacht race.[23] Its location is now known as the Eastern Garbage Patch and is between Hawaii and California. The GPGP is two distinct collections of debris bounded by the massive North Pacific Subtropical Gyre: the Western Garbage Patch near Japan and the Eastern Garbage Patch near California (Figure 2.8).[24]

The debris is spread across the water's surface and from the surface to the ocean floor. The debris ranges in size from large abandoned fishing nets to microplastics, which are not immediately noticeable to the naked eye. Sailing through areas of a garbage patch and seeing very little to no debris is possible. Garbage patches are constantly moving with ocean currents and winds. Estimating its size is challenging due to its ever-changing nature. However, the patch is estimated to cover an area of approximately 1.6 million square kilometers, three times the size of France. It is thought to contain over 1.8 trillion pieces of plastic, weighing about 80,000 metric tons, a quantity that continues to grow.

The primary cause of the GPGP is the massive amount of plastic waste that humans generate daily, with an estimated 8–15 million tons of plastic garbage released into the ocean each year. Single-use plastics and abandoned and lost fishing gear significantly contribute to the GPGP.

The GPGP is an alarming environmental issue that has gained significant attention recently. Over the decades, this massive accumulation of plastic and other debris floating in the Pacific Ocean has grown, endangering marine life and threatening the planet's health.

2.3.3 COASTS AND BEACHES

A study from the University of Bern provides new insights into plastic waste pollution in the world's oceans.[25] Models based on ocean currents have suggested that the

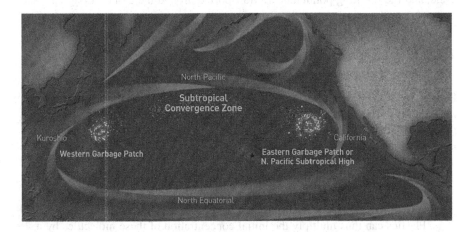

FIGURE 2.8 The Great Pacific Garbage Patch comprises the Western Garbage Patch near Japan and the Eastern Garbage Patch between Hawaii and California. It is formed by the North Pacific Gyre (National Oceanic and Atmospheric Administration, 2023).

plastic mainly collects in large ocean gyres. Now, the Bernese modelers have calcu-
lated the distribution of plastic waste on a global scale while taking into account the
fact that plastic can get beached. Their study, published in *Environmental Research
Letters*, concluded that most of the plastic does not end up in the open sea. Far more
than previously thought remains near the coast or ends up on beaches. Regions where
an above-average proportion of plastic escapes to the open sea were also identified.
In these areas, to prevent plastic from reaching the open ocean, fishing plastic out of
large rivers or removing plastic from coastlines might be more effective than cleaning up the sea.

2.4 ECOLOGICAL AND BIOLOGICAL IMPACTS

Since 1950, approximately 9.2 billion tonnes of plastic have been produced,
generating approximately 6.9 billion tonnes of primary plastic waste.[26,27] Plastic
pollution significantly impacts ecosystems, economies, and society—including
human health. Plastic pollution can alter habitats and natural processes, reducing
ecosystems' ability to adapt to climate change and directly affecting millions of
people's livelihoods, food production capabilities, and social well-being. Plastics' environmental, social, economic, and health risks must be assessed alongside
other environmental stressors, like climate change, ecosystem degradation, and
resource use.

2.4.1 PLASTICS AND BIODIVERSITY

Given the persistent nature of plastic and its toxicity, plastic pollution is a significant threat to biodiversity.[28] It threatens ecosystems and animal and plant species,
impeding their ability to deliver essential services to humanity. Indeed, plastic and
chemical leakage into the environment may arise at various stages of the plastic life
cycle, and the resulting pollutants are transported around the globe through air and
ocean streams.

2.4.2 PLASTICS AND CHEMICAL POLLUTION, INCLUDING
IMPACT ON THE FOOD CHAIN

Plastic is a viable transporter of toxic additives and pathogens in the environment,
including the food chain.[29,30] Microplastics can move through the food chain along a
process called the "trophic transfer" of microplastics. There are three main processes
through which pollution is transported:

- The first is that **pollutants present in the environment** attach themselves
  to plastic, much like iron to a magnet. Some plastics can concentrate pollutants in the environment during their stays in rivers, streams, and oceans.[31]
  Plastics can thus multiply the initial concentration of these molecules by a
  factor of up to 100,000. These molecules may also bioaccumulate in living
  organisms, i.e., concentrate along the food chain.

- Then, some chemicals **are added** to plastics during production to give them the desired properties, and these can leak out of the plastic, even when that plastic is inside an animal's body.
- Finally, plastics can remain in the environment for a long time, acting as vectors for **pathogens**.[32] In this way, plastics can spread microorganisms, including pathogens, in the sea and the air. They have even given their name to a new marine microbial habitat/ecosystem called the *"plastisphere,"* which can also be defined as the layer of microbial life that forms around a floating plastic.

2.4.3 PLASTICS AND ECOSYSTEMS, INCLUDING CORAL REEFS

Plastic pollution is damaging ecosystems around the world.[33] Plastic pollution has entered the marine food web and has significantly affected the productivity of some of the world's most important marine ecosystems, like coral reefs and mangroves.[34] The world's largest mangrove forest is becoming a plastic cesspit.[35] Root and sediment within mangroves are efficient at trapping plastics.

Plastic trash increases the risk of diseases for coral reefs.[36] Coral reefs provide vital fisheries and coastal defense, and they urgently need protection from the damaging effects of plastic waste.

Lamb et al.[37] surveyed 159 coral reefs in the Asia-Pacific region. Billions of plastic items were entangled in the reefs. The spikier the coral species, the more likely they were to snag plastic. Disease likelihood increased 20-fold once a coral was draped in plastic. Plastic debris stresses coral through light deprivation, toxin release, and anoxia, giving pathogens carried by plastics a base for invasion.

Although recognized as a global concern, the distribution of plastics trapped in the world's coral reefs remains uncertain. Pinheiro et al.[38] surveyed 84 shallow and deep coral ecosystems for anthropogenic macro-debris at 25 locations across the Pacific, Atlantic, and Indian Ocean basins. Their results show anthropogenic debris in 77 of the 84 reefs surveyed, including in some of Earth's most remote and near-pristine reefs, such as uninhabited central Pacific atolls.

Macroplastics represent 88% of the anthropogenic debris and peak in deeper reefs like other debris types, with fishing activities as the primary source of plastics in most areas. These findings contrast with the global pattern observed in other nearshore marine ecosystems, where macroplastic densities decrease with depth and are dominated by consumer items.

Microplastics can harm corals in three significant ways: abrasion, ingestion, and transfer of pollutants and pathogens.[39] Corinaldesi et al.[40] investigated the effects of microplastics on red coral, a genus with near-global distribution. When exposed to microplastics, corals preferentially ingest polypropylene, with multiple biological effects, from feeding impairment to mucus production and altered gene expression. Microplastics can alter the coral microbiome by causing tissue abrasions that allow the proliferation of bacteria. These multiple effects suggest that microplastics can ultimately cause coral death. Other habitat-forming species are likely subjected to similar impacts.

2.4.4 PLASTICS AND HUMAN HEALTH

The real impact of plastic on human health is not yet known, although micro- and nano-plastic particles have been found in the human body.

Plastic pollution threatens not only the environment but can also affect our health and future generations.[41] Humans are exposed to microplastics and plastic-carried pollutants through inhalation, ingestion, and direct skin contact throughout the plastic life cycle (Figure 2.9).

According to the World Wildlife Fund (WWF), an average person could ingest approximately 5 grams of plastic weekly. The toxic additives and pollutants found in plastics threaten human health on a global scale. Health effects include causing cancer or changing hormone activity, which can lead to reproductive, growth, and cognitive impairment. Recently, it has been pointed out that microplastics can even be found in human placentas.[42] Many toxic additives persist in the environment and bioaccumulate in exposed organisms. Microplastics also act as vessels for pathogens to enter our system, increasing the spread of diseases.

2.5 ENVIRONMENT, SOCIAL, ECONOMIC, AND POLITICAL IMPLICATIONS

Plastics are the largest, most harmful, and most persistent fraction of marine litter, accounting for at least 85% of total marine waste.[43] There is a growing threat from marine litter and plastic pollution in all ecosystems from source to sea. While we have most of the know-how and growing public support, we need the government's

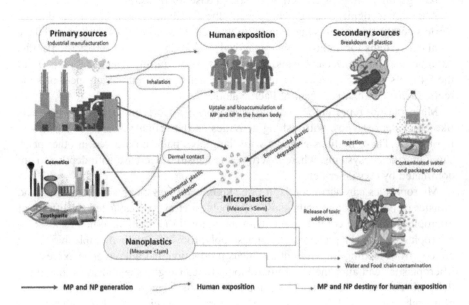

FIGURE 2.9 Microplastics (MP) and nano-plastics (NP) generation and human exposition to these plastic particles (K.L. Morales Cano et al., 2023).

political will and urgent regulatory action to tackle the problem. Politics and communicative action are vital in implementing social, cultural, and economic change toward sustainable plastics.[44] As Inger Andersen, executive director of UNEP, wrote:

> The speed at which ocean plastic pollution is capturing public attention is encouraging. We must use this momentum to focus on opportunities across the life cycle of plastics and from source to sea for clean, healthy and resilient oceans while at the same time contributing to vital Earth system processes, such as climate regulation, and to clean water, healthy ecosystems and biodiversity integrity.

The environmental and social costs of their use are significant. The annual economic costs of marine plastic pollution, concerning its impacts on tourism, fisheries, and aquaculture, and other charges, including clean-up activities, are estimated to be at least US$6–19 billion per year globally. By 2040, the expected mass of plastic leakage into the ocean will result in a US$100 billion annual financial risk for businesses if governments require them to cover waste management costs.

More than 100,000 marine studies have been conducted on the lethal and non-lethal effects of litter and plastics on every food web level, including algae, zooplankton, crustacea and invertebrates, fish, birds, turtles, and mammals. Figure 2.10 shows the extensive damage that marine litter and plastics cause to marine life and ecosystems and the potential risks to human health.

The human health impacts of marine litter and plastic pollution arise mainly from mismanaged waste handling, ingestion of seafood, and exposure to pathogenic microorganisms (Figure 2.11).

2.6 EXISTING AND IN-PROGRESS REGULATIONS

2.6.1 UNITED NATIONS

2.6.1.1 International Legally Binding Instrument

The plastic crisis has environmental, health, economic, and social impacts. We must change how we design, use, and reuse plastics, and regulation plays a role. Given that plastic affects life on Earth, the benefits of reducing plastic pollution aren't confined to one of the UN Sustainable Development Goals.

2.6.1.1.1 Intergovernmental Negotiating Committee
In March 2022, at the fifth session of the UN Environment Assembly (UNEA-5.2), the resolution 5/14 *End plastic pollution: towards an international legally binding instrument* was adopted, recognizing that "the high and rapidly increasing levels of plastic pollution represent a serious environmental problem at a global scale, negatively impacting the environmental, social and economic dimensions of sustainable development."[45,46] The resolution mandates the executive director of the UN Environment Programme (UNEP) to convene an Intergovernmental Negotiating Committee (INC) to develop the instrument based on a comprehensive approach that addresses the entire life cycle of plastic. The ambition of the INC is to complete the negotiations by the end of 2024.

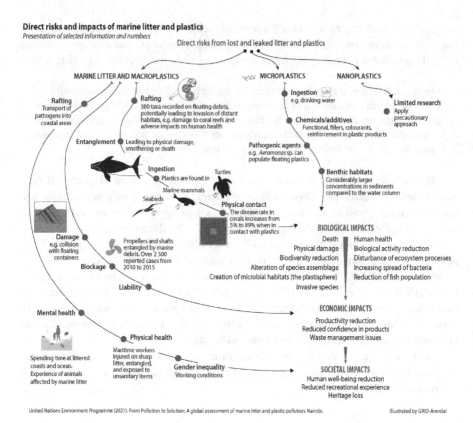

FIGURE 2.10 Direct risks and impacts of marine litter and plastics (International Union for Conservation of Nature, 2017).

The INC's first session (INC-1) occurred in Punta del Este, Uruguay, from 28 November to 2 December 2022, followed by a second session in Paris, France, from 29 May to 2 June 2023. The second session concluded with a mandate for the INC chair to prepare a zero draft of the agreement before the next session (INC-3), scheduled from 13 to 19 November 2023 at the UNEP Headquarters in Nairobi, Kenya.[47] The INC-3 delegates requested a revised zero draft for consideration at INC-4 without a single draft as its outcome.[48] The following sessions agreed upon are INC-4 in Ottawa, Canada, April–May 2024, and INC-5 in Busan, Republic of Korea, October–November 2024, followed by a Diplomatic Conference of Plenipotentiaries in mid-2025 to adopt the instrument.

2.6.1.1.2 Zero Draft of the Instrument

In September 2023, the zero-draft text of the international legally binding instrument on plastic pollution, including in the marine environment (commonly known as the Global Plastics Treaty), was released. The draft laid the foundations for the INC meeting (INC-3) in November 2023. It compiled all the suggestions made by delegations and institutions at INC-1 and INC-2 and is the base for further negotiations.

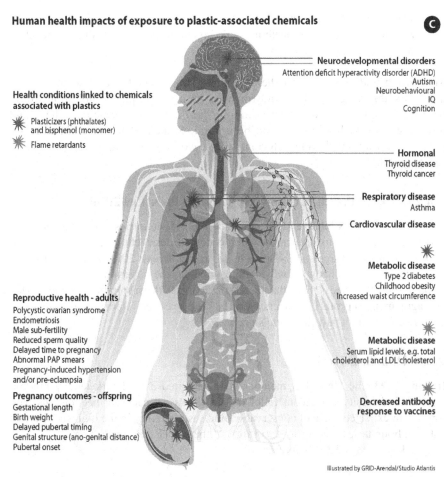

Human health impacts of exposure to plastic-associated chemicals C

Illustrated by GRID-Arendal/Studio Atlantis

United Nations Environment Programme (2021). From Pollution to Solution: A global assessment of marine litter and plastic pollution. Nairobi.

FIGURE 2.11 Human exposure to microplastic and nano-plastic particles (International Union for Conservation of Nature, 2021).

The full range of views is indicated in the draft text through up to three options for many of the obligations proposed.

Most countries support a treaty with global and legally binding rules that include bans and a robust circular economy. The draft outlines how core features (bans, phase-outs, product design requirements) can be formulated as global obligations. It contains critical measures that are necessary to curb plastic pollution. These include provisions for:

- Common binding global rules on plastic production and consumption
- Global bans or phasing out of high-risk plastics such as single-use products and intentionally added microplastics
- Product design requirements

- Waste management standards
- Extended producer responsibility covering the entire life cycle of the product
- Strong financial mechanism

Monitoring and evaluation of the agreement should use the best science available.[49] Establishing a global fund financed by a fee on plastic pollution would be a significant advance in harnessing the "polluter pays" principle. The sound governance of such a fund would be crucial. Stakeholders most impacted by plastic pollution, including informal-sector workers, should have their rights safeguarded.[50]

2.6.2 30 × 30 Pledge: Conserve 30% of the Planet by 2030

During the 2022 UN biodiversity conference, COP15, countries reached a landmark agreement that aims to reverse the unprecedented destruction of nature.[51] One of the agreement's 23 targets, 30 × 30, aims to protect at least 30% of the planet's land and water by 2030. That goal, which almost doubles the target for 2020 that was set through the UN process more than a decade ago, was the inspiration behind a 2023 UN agreement to protect biodiversity in the high seas, the international waters that comprise more than half the world's oceans.

2.6.3 High Seas Treaty

In June 2023, the UN's 193 member states adopted a landmark legally binding marine biodiversity agreement following nearly two decades of negotiations over forging an expected wave of conservation and sustainability in the high seas beyond national boundaries—covering two-thirds of the planet's oceans.[52]

Five key points of the treaty are:

- Fresh protection beyond borders
- Cleaner oceans
- Sustainably managing fish stocks
- Lowering temperatures
- Vital for realizing the 2030 Agenda, accomplishing the 17 Sustainable Development Goals by 2030

The historic treaty is crucial for enforcing the 30 × 30 pledge made by countries at the UN biodiversity conference to protect a third of the sea (and land) by 2030. Without a treaty, this target would undoubtedly fail, as until now, no legal mechanism existed to set up marine protected areas (MPAs) on the high seas. Covering almost two-thirds of the ocean that lies outside national boundaries, the treaty will provide a legal framework for establishing MPAs to protect against the loss of wildlife and share the genetic resources of the high seas. It will establish a conference of the parties that will meet periodically and enable member states to be held accountable for issues such as governance and biodiversity.

2.6.4 EUROPEAN UNION

The EU is taking action to tackle plastic pollution and marine litter to accelerate the transition to a circular and resource-efficient plastics economy.[53] Specific rules and targets apply to certain areas, including single-use plastics; plastic packaging; microplastics; and biobased, biodegradable, and compostable plastics.

As part of the circular economy action plan, the EU's plastics strategy outlines specific actions in more detail. The EU policy on plastics aims to protect the environment and human health by reducing marine litter, greenhouse gas emissions, and our dependence on imported fossil fuels. The EU also aims to:

- Transform the way plastic products are designed, produced, used, and recycled in the EU
- Transition to a sustainable plastics economy
- Support more sustainable and safer consumption and production patterns for plastics
- Create new opportunities for innovation, competitiveness, and jobs
- Spur change and set an example at the global level

Specific policies are:

- Single-use plastics: rules
- Plastic bags: rules
- Plastic packaging: rules
- Plastic waste importing and exporting shipments: rules
- Microplastics: initiative
- Global action on plastics: toward a global agreement
- Biobased, biodegradable, and compostable plastics: initiative

2.6.4.1 Plastics Strategy

The EU's plastics strategy aims to transform the way plastic products are designed, produced, used, and recycled in the EU.[54] The EU adopted a european strategy for plastics in January 2018. It is part of the EU's circular economy action plan and builds on existing measures to reduce plastic waste. The plastics strategy is critical to Europe's transition toward a carbon-neutral and circular economy. It will contribute to reaching the 2030 Sustainable Development Goals, the Paris Climate Agreement objectives, and the EU's industrial policy objectives.

The plastics strategy aims to protect our environment and reduce marine litter, greenhouse gas emissions, and our dependence on imported fossil fuels. It will support more sustainable and safer consumption and production patterns for plastics.

Actions include:

- Making recycling profitable for business
- Curbing plastic waste with a directive on single-use plastic products and fishing gear

- Driving innovation and investments
- Spurring global change with international collaboration

2.6.4.2 Single-Use Plastics

Single-use plastic products are used once or briefly before being thrown away.[55] The impacts of this plastic waste on the environment and our health are global and can be drastic. Single-use plastic products are more likely to end up in our seas than reusable options. The ten most commonly found single-use plastics represent 70% of all marine litter in the EU.

Directive (EU) 2019/904[56] on single-use plastics aims to prevent and reduce the impact of certain plastic products on the environment and promote a transition to a circular economy throughout the EU. Through the directive, different measures are being applied to other products. The ten items addressed by the directive are:

- Cotton bud sticks
- Cutlery, plates, straws, and stirrers
- Balloons and sticks for balloons
- Food containers
- Cups for beverages
- Beverage containers
- Cigarette butts
- Plastic bags
- Packets and wrappers
- Wet wipes and sanitary items

Where sustainable alternatives are readily available and affordable, single-use plastic products cannot be placed in the EU member states' markets. This applies to cotton bud sticks, cutlery, plates, straws, stirrers, and sticks for balloons. It will also apply to cups, food and beverage containers made of expanded polystyrene, and all products made of oxo-biodegradable plastic.

For other single-use plastic products, the EU is focusing on limiting their use through:

- Reducing consumption through awareness-raising measures
- Introducing design requirements, such as a requirement to connect caps to bottles
- Introducing labeling requirements
- Introducing waste management and clean-up obligations for producers, including extended producer responsibility (EPR) schemes

2.6.4.3 Plastic Bags

The Plastic Bags Directive (Directive [EU] 2015/720)[57] is an amendment to the Packaging and Packaging Waste Directive (PPWD, 94/62/EC) and was adopted to deal with the unsustainable consumption of lightweight plastic carrier bags (with a thickness below 50 microns).[58]

The directive requires member states to take measures, such as national reduction targets and/or economic instruments (e.g., fees, taxes) and marketing restrictions (bans). The measures taken by member states include either or both of the following:

- The adoption of measures ensuring that the annual consumption level does not exceed 90 lightweight plastic carrier bags per person by 31 December 2019 and 40 lightweight plastic carrier bags per person by 31 December 2025
- The adoption of instruments ensures that by 31 December 2018, lightweight plastic carrier bags are not provided free of charge at the point of sale of goods or products.

Very lightweight plastic carrier bags (with a thickness below 15 microns) may be excluded from these objectives.

Member states cannot adopt marketing restrictions (bans) for plastic carrier bags with a wall thickness above 50 microns (reusable bags). However, they are free to adopt other measures to reduce their consumption.

Since 2020, member states have reported data on the annual consumption of lightweight and very lightweight plastic carrier bags.

2.6.4.4 Packaging Waste

EU rules on packaging cover all types of packaging and packaging waste placed on the European market.[59] This includes all materials and packaging, including industrial, commercial, household, and other sectors. In 1994, the PPWD (94/62/EC) entered into force.

The PPWD aims to:

- Harmonize national measures on packaging and the management of packaging waste
- Provide a high level of environmental protection
- Ensure the excellent functioning of the internal market

The 2018 amendment to the PPWD contains updated measures to prevent the production of packaging waste and promote the reuse, recycling, and other forms of recovering packaging waste instead of its final disposal.[60] The 2018 amendment also set the following specific targets for plastic recycling: current target: 25%; by 2025: 50%; by 2030: 55%. Furthermore, by 2024, EU countries should establish producer responsibility schemes for all packaging.

In November 2022, as part of the European Green Deal and the new circular economy action plan, the Commission revised the PPWD. The initiative's objective is to ensure that all packaging is reusable or recyclable in an economically feasible way by 2030. The aim is to reinforce the essential requirements for packaging to ensure its reuse and recycling, boost the uptake of recycled content, and improve the requirements' enforceability. Measures are also envisaged to tackle over-packaging and reduce packaging waste.

2.6.4.5 Plastic Waste Shipments

In the last decade, uncontrolled trade in plastic waste has increased, damaging the environment and public health. The EU has introduced rules on the shipments of plastic waste, implementing the decision taken in 2019 at the 14th Conference of the Parties of the Basel Convention.[61]

The new rules, through Delegated Regulation (EU) 2020/2174, ban the export of plastic waste from the EU to non–Organisation for Economic Co-operation and Development (OECD) countries, except for clean plastic waste sent for recycling. Exporting plastic waste from the EU to OECD countries and imports in the EU will also be more strictly controlled. These rules should end the export of plastic waste to third countries that often do not have the capacity and standards to manage it sustainably.

2.6.4.6 Biobased, Biodegradable, and Compostable Plastics

Alternative plastics, such as biobased, biodegradable, and compostable plastics, may be a more sustainable alternative to fossil-based, non-biodegradable plastics.[62] However, they also present their sustainability challenges. In 2022, the European Commission adopted a policy framework on sourcing, labelling, and using biobased, biodegradable, and compostable plastics.[63] The communication for the EU policy framework is not legally binding. No EU law currently applies comprehensively to biobased, biodegradable, and compostable plastics. This EU policy framework aims to contribute to a **sustainable plastics economy** by improving the understanding of these materials and clarifying where these plastics can bring genuine environmental benefits and under which conditions and applications.

Biobased plastics are fully or partially made from biomass rather than fossil materials. They are not necessarily biodegradable or compostable. Biodegradable plastics biodegrade in specific conditions at the end of their life. Compostable plastics, a subset of biodegradable ones, typically decompose in industrial composting facilities. Biodegradable and compostable plastics may be made from biomass or fossil materials.

2.6.4.7 Unintentional Release of Microplastics

In the European Green Deal and new circular economy action plan, the European Commission announced a new initiative to address the unintentional release of microplastics in the environment.[64] It aims to:

- Develop labeling, standardization, certification, and regulatory measures on the unintentional release of microplastics, including measures to increase the capture of microplastics at all stages of a product's life cycle.
- Further develop and harmonize methods for measuring unintentionally released microplastics, especially from tires and textiles, and delivering harmonized data on microplastic concentrations in seawater.
- Close the gaps in scientific knowledge related to the risk and presence of microplastics in the environment, drinking water, and food.

2.6.4.8 Intentionally Added Microplastics

The deliberate use of microplastics in products such as sports pitches, detergents, diapers, and cosmetics is a significant and dangerous source of plastic pollution.[65] These microplastics end up everywhere, accumulating in the oceans and mountains. They are found in animals, food, drinking water, and the human body.

In September 2023, the Commission took a significant step to protect the environment by adopting measures that restrict microplastics intentionally added to products under the EU chemical legislation REACH.[66] The new rules will prevent the release of about half a million tonnes of microplastics into the environment. They will prohibit the sale of microplastics as such and of products to which microplastics have been added on purpose, and they will release those microplastics when used. When duly justified, derogations and transition periods for the affected parties to adjust to the new rules apply.

2.6.4.9 Global Action on Plastics

The EU is preparing the way for a new global agreement on plastics to support the shift to a circular economy, as outlined in the circular economy action plan and the mandate by the United Nations Environment Assembly (UNEA), which in 2022 launched negotiations for a new international legally binding instrument on plastic pollution.[67] The instrument should address plastic pollution throughout the entire life cycle to prevent plastic from entering the environment.

No dedicated international instrument is designed to prevent plastic pollution throughout the entire plastic life cycle. Recent studies show we can only reduce marine plastic pollution by 7% with the current measures. Plastic production is forecasted to continue growing, and more single-use plastic waste is generated than ever before.

Turkey has become the EU's top waste destination. In 2020 and 2021, the country received about half of the plastic waste that the 27 EU member states did not process in their territory, causing health and pollution problems.[68]

2.6.5 United States

2.6.5.1 Draft National Strategy to Prevent Plastic Pollution

The "Draft National Strategy to Prevent Plastic Pollution," part of a series on building a circular economy for all, builds upon the Environmental Protection Program's (EPA's) National Recycling Strategy and focuses on actions to reduce, reuse, collect, and capture plastic waste.[69]

New and innovative circular approaches are necessary to reduce and recover plastic materials and improve economic, social, environmental, and health impacts. The EPA identified three critical objectives for the strategy[70]:

- **Objective A**: Reduce pollution during plastic production.
  A1. Reduce the production and consumption of single-use, unrecyclable, or frequently littered plastic products.
  A2. Minimize pollution across the life cycle of plastic products

- **Objective B**: Improve post-use materials management.
  B1. Conduct a study of the effectiveness of existing public policies and incentives upon the reuse, collection, recycling, and conservation of materials.
  B2. Develop or expand capacity to maximize the reuse of materials.
  B3. Facilitate more effective composting and degradation of certified compostable products.
  B4. Increase solid waste collection and ensure that solid waste management does not adversely impact communities, including those overburdened by pollution.
  B5. Increase public understanding of plastic mismanagement's impact and how to appropriately manage plastic products and other waste.
  B6. Explore possible ratification of the Basel Convention and encourage environmentally sound management of scrap and recyclables traded with other countries.
- **Objective C:** Prevent trash and micro/nano-plastics from entering waterways and remove escaped trash from the environment.
  C1. Identify and implement policies, programs, technical assistance, and compliance assurance actions that effectively prevent trash/ micro/nano-plastics from getting into waterways or remove such waste once it is there.
  C2. Improve water management to increase trash and micro/ nano-plastic capture in waterways and stormwater/wastewater systems.
  C3 Increase and improve measurement of trash loadings into waterways to inform management interventions
  C4. Increase public awareness of the impacts of plastic products and other types of trash in waterways
  C5 Increase and coordinate research on micro/nano-plastics in waterways and oceans.

2.6.5.2 National Recycling Goal and National Recycling Strategy

In 2020, the EPA announced the "National Recycling Goal" to increase the U.S. recycling rate to 50% by 2030.[71]

This goal will help interested parties learn how the United States manages materials more sustainably. It will allow governments to make necessary changes to collection and sorting systems, help the industry determine the supply of available materials, and make investment decisions.

"The National Recycling Strategy: Part One of a Series on Building a Circular Economy for All," announced in 2021, is focused on enhancing and advancing the national municipal solid waste (MSW) recycling system and identifies strategic objectives and stakeholder-led actions to create a stronger, more resilient, and cost-effective domestic MSW recycling system.[72,73]

The strategy responds to the U.S. recycling system's challenges through actions outlined under five objectives:

- Improve markets for recycled commodities
- Increase collection and improve materials management infrastructure
- Reduce contamination in the recycled materials stream

- Enhance policies and programs to support circularity
- Standardize measurement and increase data collection

2.6.6 CHINA

China has been contending with plastic pollution for years. China's severe and concentrated effort to govern plastics took off in 2016.[74] Since then, China has experienced a notable increase in attention dedicated to addressing plastic pollution through regulatory measures. In 2000, only four policies were explicitly focused on plastic-related issues. However, by the first half of 2021, this number had surged to 41, marking a remarkable increase of 925%. In 2020, China announced a phased ban on single-use plastics. Nondegradable bags were prohibited in major cities by the end of 2020 and in all cities and towns by 2022.[75] Additionally, China has initiated measures to restrict the importation of recyclable materials into the country as an additional strategy to combat plastic pollution.[76]

However, considering the magnitude of the challenge, China must undertake more extensive actions than those currently implemented. China's ban on the importation of plastic waste has merely shifted the problem to other regions and countries. With China effectively closing its doors to plastic waste imports, hundreds of small-scale Chinese plastic recyclers have relocated to other Southeast Asian countries, particularly Malaysia, Vietnam, and Thailand.[77]

Since 2016, the Chinese government has intensified its efforts to combat plastic pollution through the development of numerous policies at both the national and local levels. However, the tangible impact of these policies remains unclear. In 2018, China disposed of 200.7 million cubic meters of waste into its coastal waters, marking a 27% increase from the previous year and reaching the highest level in at least a decade, according to the country's environment ministry.

Most of this waste was deposited in the Yangtze and Pearl River delta regions, major industrial zones on China's eastern coast.[78] To maintain a positive trajectory, the new policies implemented in China could serve as a model for other nations in addressing the issue of plastic pollution. They have the potential to contribute to the advancement of a global plastic pollution treaty, as emphasized by a WWF official during the third session of the UN's INC (INC-3).[79]

2.6.7 INDIA

India has been taking steps to reduce plastic waste for several years.[80] India generates 3.5 million tons of plastic waste annually, according to the environment minister, and remains one of the biggest plastic polluters in the world.[81] In July 2022, India imposed a ban on single-use plastics to tackle the country's rapidly increasing levels of plastic pollution.[82] The ban includes straws, cutlery, earbuds, packaging films, and cigarette packets, among other products.[83] Companies and plastic manufacturers have complained about the ban, lobbying for items to be removed and saying they were not given adequate time to prepare.

India ranks among the top producers of plastic waste globally, and the widespread production and consumption of plastic items have resulted in the proliferation of microplastics, an emerging class of environmental contaminants. There exists a growing concern surrounding microplastic pollution across various ecological compartments. A limited number of studies have been conducted in India to understand the impact of microplastic pollution on aquatic systems, terrestrial systems, the atmosphere, and human consumption.[84]

The National Green Tribunal in India has acknowledged a report highlighting the potential for microplastics to infiltrate blood cells and contaminate drinking water sources. One of India's primary contributors to microplastic pollution is the widespread use of plastic bags.[85]

The volume of plastic waste generated in India is not the primary concern; instead, the inadequate management of this waste poses a significant challenge. Plastic waste in India is often inefficiently collected, managed, disposed of, and recycled. According to Recykal, a tech startup offering digital solutions for sustainability, the demand for plastic in India soared to approximately 21 million tons in 2021. However, the recycling rate remains alarmingly low, with India managing to recycle only 12% of its plastic waste thus far. All plastic manufacturing and recycling units in India must be registered with the relevant State Pollution Control Boards (SPCBs) or Pollution Control Committees (PCCs). Currently, there are 4953 registered units across the country.[86]

At the fifth UNEA held in Nairobi in 2022, India actively engaged with member states to establish a consensus on a resolution to drive global action against plastic pollution. This resolution proposed the establishment of an intergovernmental negotiating committee to develop a new international legally binding instrument.[87]

However, it is concerning to note that India, along with countries with significant petrochemical interests, has worked to weaken critical aspects of the proposed Global Plastics Treaty despite clear scientific evidence of harm and the vulnerability of its citizens. The Global Plastics Treaty represents an unprecedented opportunity for coordinated global action against plastic pollution. Nevertheless, significant challenges persist, exacerbated by the conflicting interests of nations and industries vested in the current status quo.[88]

The Indian government's alignment with nations like Saudi Arabia, Russia, China, the United States, and Iran—each with substantial plastic-petrochemical interests— has led to the insistence that the treaty should primarily focus on downstream waste management. This stance disregards the scientific evidence surrounding climate change, plastic pollution, and the destabilization of Earth's systems.[89]

2.6.8 JAPAN

Japan is known for its excessive use of plastic, being the second-highest consumer of single-use plastic in the world.[90] The country's plastic recycling rate is only around 27.8%, and more than 50% of plastic waste is converted back into fuel, with a further 14% incinerated with other waste.[91] This has led to significant plastic waste in the surrounding oceans, affecting marine life and ecosystems. By 2030, Japan aims to reuse or recycle 60% of all plastic containers and packaging, reduce single-use plastic emissions by a quarter, and introduce fewer polluting bioplastics. By 2035, Japan aims to reuse or recycle all plastic waste.[92]

Like many other countries, Japan has become increasingly aware of the environmental problems caused by plastics in the oceans. Efforts are being made to combat this problem and to raise public awareness through various channels.

The Japanese government has taken steps to tackle plastic pollution. For example, regulations and initiatives have been implemented to reduce single-use plastics. Policies have also been implemented to encourage recycling and proper disposal of plastics. The "MARINE Initiative," which focuses on waste management, marine debris recovery, innovation, and empowerment, has been launched.[93]

In addition to its work on plastic waste, the Japanese government, through the Japan Initiative for Marine Environment, demonstrates Japan's international cooperation efforts to improve plastic waste management in Asia.[94] Japan's influence varies depending on multiple factors, including diplomatic relations, economic ties, and the willingness of neighboring countries to adopt new practices; its efforts and contributions to technology, international collaboration, aid, culture, and education can collectively play a pivotal role in extending public awareness about plastics in the oceans to its neighboring countries in Southeast Asia. International activities by Japanese companies, nongovernmental organizations (NGOs), and local governments facilitate the export of infrastructure, such as waste management facilities, and introduce innovation and technology regarding plastic alternatives and recycling.[95]

In Japan, the dissemination and sharing of best practices on waste in the public and private sectors has continued through various channels. Several companies have also taken steps to reduce their plastic footprint. They have implemented plastic reduction policies, switched to environmentally friendly packaging, or launched initiatives to promote recycling. The media, including television, newspapers, and online platforms, often cover stories related to environmental issues, including plastic pollution in the oceans. Documentaries and news reports highlight the consequences of plastic waste and its impact on marine life.

Schools and educational institutions are incorporating environmental studies into their curriculum, including lessons on the impact of plastic pollution on the oceans. This helps instill awareness and responsibility for reducing plastic waste in the younger generation. Projects such as Umi to Nihon (The Sea and Japan) educate people about ocean-related issues. Groups organize awareness campaigns, beach clean-ups, and other initiatives to engage the public and highlight the impact of plastics on marine ecosystems. Through neighborhood associations or volunteer groups, local communities often organize clean-ups along rivers, beaches, and coastal areas to remove plastic waste and raise awareness.

2.6.9 SOUTHEAST ASIA

Southeast Asia is one of the regions most affected by marine plastic pollution, as it produces about half of the world's mismanaged plastic waste. However, various efforts and initiatives are also underway to raise public awareness and reduce plastic consumption and waste in the region.

The Association of Southeast Asian Nations (ASEAN) member states have adopted a Regional Action Plan to tackle plastic pollution, which includes 14 actions across four pillars: policy support and planning; research, innovation, and capacity building; public awareness, education, and outreach; and private-sector engagement.[96] The plan aims to enhance regional cooperation and coordination and support national and local actions to address the issue. Several UN agencies, such as Economic and Social Commission for Asia and the Pacific (ESCAP) and UNEP, have launched projects and campaigns to help Southeast Asian countries reduce plastic pollution and its impacts on marine ecosystems. For example, the "Closing the Loop" project aims to use innovative technologies, such as remote sensing and satellite data, to detect and monitor the sources and pathways of plastic waste in urban areas. The project will be piloted in four ASEAN cities: Kuala Lumpur, Surabaya, Nakhon Si Thammasat, and Da Nang.

Several NGOs, individuals, and public and private organizations are also working to raise public awareness and promote behavioral change toward plastic consumption and waste in Southeast Asia. For example, the Plastic Smart Cities initiative supports cities and tourism destinations in reducing plastic pollution by providing tools and resources to implement plastic waste management strategies. The initiative has engaged more than 40 regional cities and destinations, such as Bali, Phuket, Siem Reap, and Hoi An. While these regional initiatives pertain to international agreements and collaborations, national initiatives are taken at different levels: (1) legislation and policy changes; (2) plastic bans and restrictions; (3) the promotion of recycling and a circular economy; (4) cleanup and awareness campaigns; (5) innovation and research; (6) corporate initiatives; and (7) consumer awareness and behavioral changes, educational programs, media coverage, community engagement, and advocacy.

Indonesia, the world's second-largest plastic waste producer, faces a significant challenge with marine plastic pollution. The country generates about 9 million tonnes of plastic waste annually, of which about 15% end up in the surrounding oceans, affecting marine life and ecosystems.[97] Indonesia has become aware of the environmental issues related to plastic in the oceans and has implemented several measures and initiatives to combat this problem and raise public awareness. The Indonesian government has adopted a national action plan to reduce marine plastic waste by 70% by 2025. This plan includes reducing single-use plastic consumption, improving plastic waste management, strengthening regulation and monitoring, developing innovation and research, and promoting regional and international cooperation.[98] In Indonesia, there is a general effort to reduce plastic pollution by educating schoolchildren, cleaning beaches, and advocating for better waste management. For example, the Pulau Plastik (Plastic Island) campaign has produced videos and a documentary to raise public awareness of the danger of plastic waste and the possible solutions. A survey among the viewers showed that the film increased awareness that the leading solution to address this crisis is reducing the usage of single-use plastic in the first place.[99] Several Indonesian companies have also taken initiatives to decrease their plastic footprint, such as adopting policies to reduce plastic, switching to eco-friendly packaging, or launching recycling programs. For example, the e-commerce company Tokopedia has established a program called "Greener Packaging" that allows sellers and buyers to choose a package made of recycled paper or biodegradable corn starch.[100]

2.6.10 EFFECTIVENESS AND LIMITATIONS OF CURRENT REGULATIONS

There has been a steady increase in awareness of the environmental, economic, and social risks posed by (micro)plastic pollution.[101] This has led to the development of numerous national, regional, and international regulatory tools. Internationally, the developed efforts vary in scope and range, focusing on manufacturing, commercializing, and using (micro)plastics. At the same time, at the national and regional levels, most initiatives endeavor to curtail plastic pollution by imposing either levies or bans. However, it is not clear what the benefits of the multitude of norms, regulations, laws, and recommendations that have been proposed and/or implemented in recent years are. The industry continues to fight some of the legislative initiatives. At the research level, there is also the need to gather more data regarding the effects of these materials to allow the development of adequate regulations.

All regulatory instruments can be considered to address pollution through the following actions:

- Preventive—focuses on the 3R rule: reuse, reduction (at sources), and recycling
- Removal—debris monitoring and clean-up initiatives
- Mitigation—litter disposal and discharge regulations
- Educational—awareness campaigns and incentive approaches

Such strategies often suffer from limited coordination between all stakeholders.

The best approach to dealing with plastic pollution will be multi-tiered, including bottom-up governance and local, national, regional, and international hard and soft laws. Corporations will have to reconsider the design of their products. Consumers will also have to adjust their behaviors and, together with producers, shift toward a culture of reduction, reuse, and recycling.

2.7 RESPONSIBILITIES OF STAKEHOLDERS

The plastic life cycle involves multiple stakeholders with different roles, responsibilities, and accountabilities.[102] In general, stakeholders are grouped along the life cycle of plastic: producers, governments, consumers, end-of-life waste management services, and civil society. Different stakeholders must not shift responsibilities onto other stakeholders, like producers or customers.

Plastic producers play a significant role in the plastic life cycle, as they determine the design, composition, and production processes of plastic products.

Governments also play a critical role in regulating the production and use of plastic. They can impose taxes, bans, or regulations on plastic production, use, and disposal. Governments can also incentivize sustainable plastic production practices and invest in waste management infrastructure to reduce plastic pollution.

Consumers are also important stakeholders in the plastic life cycle, as they use and dispose of plastic products. They can influence the demand for sustainable products, recycle plastic waste, or reduce plastic consumption altogether.

Waste management companies collect, sort, and dispose of plastic waste. They can provide recycling services and educate consumers on proper waste management practices.

Civil society (e.g., NGOs, trade and industry associations) is also an essential stakeholder in the plastic life cycle, as it can advocate for policy changes, raise awareness of plastic pollution, and promote sustainable consumer behavior.

2.8 UNEP PROPOSAL FOR A SYSTEMS CHANGE: "TURNING OFF THE TAP"

In parallel with the INC tackling accountability from manufacturing to disposal, the UNEP has released a report titled "Turning Off the Tap, How the World Can End Plastic Pollution and Create a Circular Economy."[103] This report proposes a systems change to address the causes of plastic pollution. The systems change scenario combines reducing unnecessary and problematic plastic uses with a market transformation toward circularity in plastics by accelerating three fundamental market shifts—reuse, recycle, and reorient and Diversify—and actions to deal with the plastic pollution legacy through the formulation of sound policies and regulatory framework (Figure 2.12).

Reuse refers to transforming the "throwaway economy" into a "reuse society"; recycle refers to recyclable plastic products; and reorient and diversify refers to shifting the market toward sustainable alternatives (Figure 2.13).

The highest costs in both a throwaway and circular economy are operational. EPR schemes can cover these operational costs of ensuring the system's circularity by requiring producers to finance the collection, recycling, and responsible end-of-life disposal of plastic products. EPR schemes have been applied to many sectors and products.[104] It is a concept where producers bear a significant degree of responsibility for the environmental impacts of their products throughout the product life cycle. EPR schemes are a practical implementation of the "polluter

FIGURE 2.12 The systems change toward a new circular plastics economy (International Union for Conservation of Nature, Turning Off the Tap, 2023).

pays" principle. The widespread adoption of EPR across the plastics economy is among the highest policy priorities for achieving circularity targets. EPR schemes are predominantly used for collection and recycling, but their expansion to broader goals, including redesign toward reduction and reuse, should be the next step.

The UNEP proposal also addresses specific policies, including standards for design, safety, and compostable and biodegradable plastics; targets for minimum recycling; EPR schemes; taxes; bans; communication strategies; public procurement; and labeling.[105]

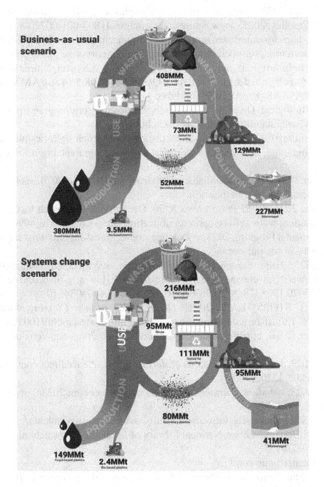

FIGURE 2.13 The short-lived plastic flows in the economy in 2040 in a business-as-usual linear economy (top) versus that projected in the systems change scenario (bottom). Under the systems change scenario, the inflow of new material into the economy of short-lived plastics is more than halved by increasing the reused or recycled flows into the economy. As a result, the outflow of mismanaged plastic waste ending in the environment decreases by over 80% (41 million tonnes instead of 227 million tonnes) (International Union for Conservation of Nature, Turning Off the Tap, 2023).

NOTES

1 J. Boucher, 2017, *Primary Microplastics in the Oceans, a Global Evaluation of Sources*, International Union for Conservation of Nature, https://portals.iucn.org/library/node/46622

2 https://ecostandard.org/reduce-reuse-redesign-what-place-for-plastic-in-a-circular-economy/

3 T. Yang, 2023, *Science of the Total Environment*, www.sciencedirect.com/science/article/pii/S0048969722078615

4 Horiba, *Primary Microplastics*, www.horiba.com/usa/scientific/resources/science-in-action/where-do-microplastics-come-from/

5 Condor Limited, 2023, www.condorferries.co.uk/plastic-in-the-ocean-statistics

6 C. Morales Caselles, 2021, www.nature.com/articles/s41893-021-00720-8

7 Neste, 2023, https://journeytozerostories.neste.com/mechanical-vs-chemical-recycling?utm_campaign=awareness_sem-belgium-brand-innovation_corporate_neste_neste_always-on-advertising_belgium_b2b_new-customers_sus_2023_q2&utm_source=google&utm_medium=search_paid&gclid=EAIaIQobChMIlZK554z6gAMVcZKDBx0wMQKyEAAYASAAEgK-jvD_BwE#79fd9bec

8 Center for Biological Diversity, www.biologicaldiversity.org/programs/population_and_sustainability/sustainability/plastic_bag_facts.html

9 European Commission, https://environment.ec.europa.eu/topics/plastics/plastic-bags_en

10 National Geographic, 2023, https://education.nationalgeographic.org/resource/one-bottle-time/

11 The Guardian, 2022, www.theguardian.com/environment/2022/nov/07/invisible-killer-ghost-fishing-gear-deadliest-marine-plastic

12 J. Toyoshima, 2021, *OPRI Perspectives N°20*, Ocean Policy Research Institute, Sasakawa Peace Foundation, www.spf.org/opri/en/publication/perspectives/?idx_409035=2

13 UN Environmental Program, 2023, www.unep.org/technical-highlight/fishing-nets-double-edged-plastic-swords-our-ocean

14 Euractiv, 2023, *From Europe to Africa and Asia*, www.euractiv.com/section/circular-economy/news/from-europe-to-africa-and-asia-the-journey-of-discarded-textiles/?_ga=2.178850020.1965378581.1696934610-997443958.1696934610

15 ABC NEWS STORY LAB, 2021, *Dead White Man's Clothes*, www.abc.net.au/news/2021-08-12/fast-fashion-turning-parts-ghana-into-toxic-landfill/100358702

16 Eco Gardener, 2018, https://ecogardener.com/blogs/news/pros-and-cons-of-using-plastic-mulch

17 N. Khalid, 2023, *Journal of Hazardous Materials*, www.sciencedirect.com/science/article/abs/pii/S030438942202249X

18 L. Beck, 2023, *Modern Farmer*, https://modernfarmer.com/2023/08/plastic-mulch-is-problematic/

19 Washington State University, https://smallfruits.wsu.edu/plastic-mulches/

20 E. Watt, 2021, *RSC Advances*, National Library of Medicine, www.ncbi.nlm.nih.gov/pmc/articles/PMC9034135/

21 https://theoceancleanup.com/

22 National Oceanic and Atmospheric Administration, 2023, *Garbage Patches*, https://marinedebris.noaa.gov/info/patch.html

23 Plastic Collection Blogger, 2023, www.plasticcollective.co/what-is-the-great-pacific-garbage-patch/

24 https://education.nationalgeographic.org/resource/great-pacific-garbage-patch/

25 University of Bern, 2021, "Plastic Waste in the Sea Mainly Drifts Near the Coast", *Environmental Research Letters*, Vol. 16, No. 6, www.sciencedaily.com/releases/2021/06/210602130246.htm

26 UN Environment Programme, 2023, https://leap.unep.org/content/basic-page/plastics-pollution-toolkit-about

27 UN Environment Programme, 2023, www.unep.org/plastic-pollution

28 Geneva Environment Network, www.genevaenvironmentnetwork.org/resources/updates/plastics-and-biodiversity/

29 Plastic Soup Foundation, www.plasticsoupfoundation.org/en/plastic-problem/what-is-plastic/plastic-pollutants/

30 Plastic Soup Foundation, www.plasticsoupfoundation.org/en/plastic-problem/plastic-affect-animals/plastic-food-chain/

31 A. Ter Halle, 2019, *Encyclopedia of Environment*, www.encyclopedie-environnement.org/en/water/plastic-pollution-at-sea-seventh-continent/

32 Plastic Soup Foundation, www.plasticsoupfoundation.org/en/plastic-problem/health/pathogens/

33 www.nature.com/articles/d41586-023-02252-x

34 L. Parker, 2019, *National Geographic*, www.nationalgeographic.com/environment/article/plastic-pollution

35 T.M. Adyel, 2021, *Frontiers*, www.frontiersin.org/articles/10.3389/fmars.2021.766876/full

36 E. Stokstad, 2018, *Science*, www.science.org/content/article/plastic-trash-making-coral-reefs-sick

37 J.B. Lamb, 2018, *Science*, www.science.org/doi/10.1126/science.aar3320

38 H.T. Pinheiro, 2023, *Nature*, www.nature.com/articles/s41586-023-06113-5

39 https://kids.frontiersin.org/articles/10.3389/frym.2021.574637

40 C. Corinaldesi, 2021, *Communications Biology*, NCBI, www.ncbi.nlm.nih.gov/pmc/articles/PMC8010021/

41 www.genevaenvironmentnetwork.org/resources/updates/plastics-and-health/

42 K.L. Morales Cano, et al., 2023, *Advances and Challenges in Microplastics, Micro(Nano) Plastics as Carriers of Toxic Agents and Their Impact on Human Health*, https://doi.org/10.5772/intechopen.111889; www.intechopen.com/online-first/87700

43 UNEP, 2021, *From Pollution to Solution: A Global Assessment of Marine Litter and Plastic Pollution*, www.unep.org/resources/pollution-solution-global-assessment-marine-litter-and-plastic-pollution

44 T.A. Farrelly, 2021, *Book, Plastic Legacies: Pollution, Persistence, and Politics*, www.researchgate.net/publication/363295935_Plastic_Legacies_Pollution_Persistence_and_Politics

45 UNEP, 2023, *Intergovernmental Negotiating Committee on Plastic Pollution*, www.unep.org/inc-plastic-pollution

46 Geneva Environment network, 2023, www.genevaenvironmentnetwork.org/fr/ressources/nouvelles/towards-plastic-pollution-inc-2/

47 UNEP, 2023, www.unep.org/news-and-stories/press-release/inc-chair-prepare-zero-draft-international-agreement-plastic

48 IISD, 2023, *INC-3 Summary*, https://enb.iisd.org/plastic-pollution-marine-environment-negotiating-committee-inc3-summary

49 The Circular Initiative, 2023, *Who We Are*, www.thecirculateinitiative.org/our-story

50 Circulate Initiative, 2023, *Responsible Sourcing*, www.thecirculateinitiative.org/responsible-sourcing

51 New York Times, 2023, *UN 2022 Biodiversity Conference, COP15*, www.nytimes.com/2022/12/19/climate/biodiversity-cop15-montreal-30x30.html

52 UN, 2023, *High Seas Treaty*, https://news.un.org/en/story/2023/06/1137857

53 European Commission, 2023, *Plastics*, https://environment.ec.europa.eu/topics/plastics_en

54 European Commission, 2018, *Environment, Plastics Strategy*, https://environment.ec.europa.eu/strategy/plastics-strategy_en

55 European Commission, 2023, *Single-Use Plastics*, https://environment.ec.europa.eu/topics/plastics/single-use-plastics_en

56 EUR-Lex, 2022, *Directive (EU) 2019/904*, https://eur-lex.europa.eu/eli/dir/2019/904/oj

57 EUR-Lex, 2015, *Plastic Bags*, https://eur-lex.europa.eu/legal-content/EN/TXT/?qid=1601561123103&uri=CELEX:32015L0720

58 European Commission, *Plastic Bags*, https://environment.ec.europa.eu/topics/plastics/plastic-bags_en

59 European Commission, 1994, *Packaging Waste*, https://environment.ec.europa.eu/topics/waste-and-recycling/packaging-waste_en

60 EUR-Lex, 2018, *Packaging Waste*, https://eur-lex.europa.eu/legal-content/EN/TXT/?uri=celex%3A32018L0852

61 European Commission, 2021, *Plastic Waste Shipment*, https://environment.ec.europa.eu/topics/waste-and-recycling/waste-shipments/plastic-waste-shipments_en

62 European Commission, 2022, *Biobased, Biodegradable and Compostable Plastics*, https://environment.ec.europa.eu/topics/plastics/biobased-biodegradable-and-compostable-plastics_en

63 EuropeanCommission,2022,*Communication*,https://environment.ec.europa.eu/publications/communication-eu-policy-framework-biobased-biodegradable-and-compostable-plastics_en

64 European Commission, 2022, *Microplastics*, https://environment.ec.europa.eu/topics/plastics/microplastics_en

65 Rethink Plastic, 2023, *European Commission Finally Restricts Intentional Use of Microplastics*, https://rethinkplasticalliance.eu/news/european-commision-finally-restricts-intentional-use-of-microplastics-in-first-concrete-victory-for-ecosystems-and-human-health/

66 European Commission, 2023, *Intentionally Added Microplastics*, https://ec.europa.eu/commission/presscorner/detail/en/ip_23_4581

67 European Commission, 2022, *Global Action on Plastics*, https://environment.ec.europa.eu/topics/plastics/global-action-plastics_en#timeline

68 Le Monde, September 22, 2022, www.lemonde.fr/en/international/article/2022/09/22/in-turkey-europe-s-new-trash-can-plastic-recycling-poses-serious-risks-to-health-and-the-environment_5997828_4.html

69 Environmental Protection Agency, 2023, *Draft Strategy to Prevent Plastic Pollution*, www.epa.gov/circulareconomy/draft-national-strategy-prevent-plastic-pollution

70 Environmental Protection Agency, 2023, *Draft Strategy to Prevent Plastic Pollution; Executive Summary*, www.epa.gov/system/files/documents/2023-04/Draft_National_Strategy_to_Prevent_Plastic_Pollution_Executive_Summary.pdf

71 Environmental Protection Agency, 2023, *National Recycling Strategy*, www.epa.gov/circulareconomy/us-national-recycling-goal

72 Environmental Protection Agency, 2023, *National Recycling Strategy*, www.epa.gov/circulareconomy/national-recycling-strategy

73 Environmental Protection Agency, 2021, *National Recycling Strategy, Part One of a Series on Building a Circular Economy for All*, www.epa.gov/system/files/documents/2021-11/final-national-recycling-strategy.pdf

74 K. Fürst, 2022, "China's Regulatory Respond to Plastic Pollution: Trends and Trajectories", *Frontiers in Marine Science*, Vol. 9, www.frontiersin.org/articles/10.3389/fmars.2022.982546/full

75 BBC, News, 2020, *Single-Use Plastic: China to Ban Bags and Other Items*, www.bbc.com/news/world-asia-china-51171491

76 P.A. Davies, 2018, *Environment, Land & Resources*, www.globalelr.com/2018/04/chinas-action-against-plastic-pollution/

77 Financial Times, 2018, *Why the World's Recycling System Stopped Working*, www.ft.com/content/360e2524-d71a-11e8-a854-33d6f82e62f8

78 Reuters, 2019, *China's Ocean Waste Surges 27% in 2018*, www.reuters.com/article/idUSKBN1X80SH/

79 English News, 2023, *Interview: China's Plastic Measures Set Standard, Will Contribute to Global Treaty, WWF Official Says*, https://english.news.cn/20231113/fca9e539e-51b41ab982292d3bad9576c/c.html

80 The Economic Times, 2017, https://economictimes.indiatimes.com/news/politics-and-nation/how-indias-efforts-towards-reducing-its-plastic-footprint-have-worked-so-far/articleshow/71415985.cms

81 The Economic Times, 2022, *India Generates 3.5 Million Tonnes Plastic Waste Annually*, https://economictimes.indiatimes.com/industry/environment/india-generates-3-5-million-tonnes-plastic-waste-annually-environment-minister/articleshow/90668558.cms?from=mdr

82 IndiaBriefing,2022,www.india-briefing.com/news/india-new-plastic-waste-management-rules-single-use-plastic-ban-effective-from-july-1-2022-25398.html/

83 CNN, 2022, https://edition.cnn.com/2022/07/01/india/india-bans-single-use-plastic-intl-hnk/index.html

84 Springer, 2021, *Microplastics as Contaminants in Indian Environment: A Review*, https://link.springer.com/article/10.1007/s11356-021-16827-6

85 The Times of India, 2023, *India and Microplastics: Redefining India's One of the Most Crucial Environmental Problems*, https://timesofindia.indiatimes.com/blogs/scientifically-trended/india-and-microplastics-redefining-indias-one-of-the-most-crucial-environmental-problems/

86 The Energy and Resources Institute, 2023, *Towards a Circular Plastics Economy: India's Actions to #BeatPlasticPollution*, www.teriin.org/article/towards-circular-plastics-economy-indias-actions-beatplasticpollution

87 Hindustan Times, 2023, *India is Moving Away from Single-Use Plastics*, www.hindustantimes.com/opinion/indias-successful-journey-towards-beating-plastic-pollution-a-year-of-bans-alternatives-and-ecoinnovation-101688136783729.html

88 United Nations, *Nations Agree to End Plastic Pollution*, www.un.org/en/climatechange/nations-agree-end-plastic-pollution

89 The Third Pole, *Opinion: Why is India Weakening the Global Plastics Treaty?* www.thethirdpole.net/en/pollution/opinion-why-is-india-weakening-the-global-plastics-treaty/

90 https://zenbird.media/marine-plastic-waste-in-japan-and-how-we-can-stop-it/

91 https://japantoday.com/category/features/opinions/japan%E2%80%99s-plastic-addiction-is-affecting-oceans-and-burdening-marine-lifejapandtoday.com

92 https://hk.boell.org/en/2022/05/30/plastic-atlas-japan-special-edition-closer-look-japans-plastic-waste-management

93 www.mofa.go.jp/ic/ge/page25e_000317.html

94 www.mofa.go.jp/files/000419527.pdf

95 www.breakfreefromplastic.org/2022/09/23/japans-plastic-waste-exports-and-how-to-slow-them-down/

96 www.ukri.org/news/impacts-of-marine-plastic-pollution-in-south-east-asia-researched/

97 www.worldbank.org/en/news/feature/2020/10/06/stemming-the-plastics-tide-in-indonesia

98 https://theconversation.com/pulau-plastik-campaign-has-raised-public-awareness-on-plastic-waste-yet-challenges-remain-167815

99 https://theconversation.com/how-can-indonesia-win-against-plastic-pollution-80966

100 www.weforum.org/press/2020/04/indonesia-unveils-action-plan-to-prevent-16-million-tonnes-of-plastic-from-entering-the-ocean/

101 J.P. da Costa, 2020, *Frontiers*, www.frontiersin.org/articles/10.3389/fenvs.2020.00104/full

102 J. Andersen, 2023, *Plastic Oceans International*, https://plasticoceans.org/solving-plastic-pollution-unequal-stakeholders/

103 UNEP, 2023, *Turning Off the Tap: How the World Can End Plastic Pollution and Create a Circular Economy*, www.unep.org/resources/turning-off-tap-end-plastic-pollution-create-circular-economy

104 UNEP, 2023, *Topic Sheet: Extended Producer Responsibility*, https://wedocs.unep.org/handle/20.500.11822/42235

105 UNEP, 2023, *Science Policy Business Forum*, https://un-spbf.org/turning-off-the-tap-how-the-world-can-end-plastic-pollution-and-create-a-circular-economy/

3 Combating Plastic Pollution

3.1 INTRODUCTION

Plastic pollution constitutes a global problem with environmental and biological impacts on marine and human life, including ecosystems, biodiversity, the climate, and human health. This, in turn, has social, economic, and political implications. Without new measures, plastic production is projected to double in 20 years, and plastic waste leaking into the ocean is expected to triple by 2040. We cannot continue using plastics as we have, but so far, we cannot live without them. Therefore, we need (1) to develop sustainable, future-proof plastics by adopting a new, sustainable way of designing and using plastics and (2) to clean up ocean pollution.

The chapter is divided into six parts:

1. International agreements and goals
2. Recycling in favor of a circular economy
3. Alternative materials toward a circular biobased economy
4. Upstream prevention solutions
5. Public and consumer awareness, including regional, corporate, and cleanup initiatives
6. A vision for sustainable plastics

Public awareness about ocean plastic pollution has steadily increased. Efforts by environmental organizations, documentaries, and social media campaigns have contributed to raising awareness about the detrimental impact of plastic waste on marine ecosystems. Many people are now more informed about the severity of the issue and understand how discarded plastics harm aquatic life, ecosystems, and potentially human health. This increased awareness has led to more significant public support for measures such as plastic bags, recycling initiatives, beach clean-ups, and reducing single-use plastics. However, the level of awareness can vary regionally and among different demographic groups. In some areas, there might be more active participation in beach cleanups and local initiatives to reduce plastic waste. In contrast, the issue might be less prominently addressed or acknowledged in others. It is important to note that the level of public awareness and actions taken may have evolved since then due to ongoing education, advocacy efforts, and governmental policies aimed at combatting plastic pollution.

DOI: 10.1201/9781003532477-3

3.2 INTERNATIONAL AGREEMENTS AND PROPOSED GOALS

3.2.1 INTERGOVERNMENTAL NEGOTIATING COMMITTEE SESSIONS

The Intergovernmental Negotiating Committee Second Session (INC-2) considered the potential core obligations consolidated by the secretariat based on written submissions from member states (Figure 3.1).[1]

The core obligation options include control measures and voluntary approaches and relate to[2]:

- Phasing out and/or reducing the supply of, demand for, and use of primary plastic polymers
- Banning, phasing out, and/or reducing the use of problematic and avoidable plastic products
- Banning, phasing out, and/or reducing the production, consumption, and use of chemicals and polymers of concern
- Reducing microplastics
- Strengthening waste management
- Fostering design for circularity
- Encouraging "reduce, reuse, and repair" of plastic products and packaging
- Promoting the use of safe and sustainable alternatives and substitutes
- Eliminating the release and emission of plastics to water, soil, and air
- Addressing existing plastic pollution
- Facilitating a just transition, including an inclusive transition of the informal waste sector
- Protecting human health from the adverse effects of plastic pollution

Critical considerations for the growth and design of biobased polymers include their end-of-life pathway:

- Biobased polymers identical to their fossil counterparts, such as bio-PE and bio-PET, can be mechanically recycled in existing recycling streams.
- PLA undergoes slow degradation in nature over a period of 6–24 months.

FIGURE 3.1 Second session of Intergovernmental Negotiating Committee on Plastic Pollution (INC-2). (Source: UNEP.)

They can be grouped into three categories along the entire life cycle of plastics:

- Upstream solutions that focus on reducing the production and consumption of primary plastic polymers, problematic and avoidable plastic products, and chemicals of concern, as well as microplastics (i.e., narrowing the loop in a circular economy)
- Midstream solutions that foster design for circularity; encourage the reduction, reuse, and repair of plastic products and packaging; and promote sustainable alternatives (i.e., operating and slowing the loop in a circular economy)
- Downstream solutions that strengthen waste management; eliminate the release of plastics to water, soil, and air; address existing plastic pollution; facilitate a just transition; and protect human health (i.e., closing the loop in a circular economy)

Strengthening waste management (downstream solutions) and designing for circularity (midstream solutions) are options readily accepted by the countries. The critical question is how to stop plastic pollution at its sources, involving production reduction and control measures. The issue was whether to put a target in place to reduce plastic production. While environmental groups and many countries considered reducing production the most effective measure, other countries, including some significant plastics producers, still needed to be ready to take this position.

The means of implementation considered relate to possible arrangements for capacity building and technical assistance, technology transfer, and financial assistance. Options for implementation measures include national action plans, national reporting, compliance provisions, and periodic assessment and monitoring of the instrument's implementation progress and effectiveness evaluation. The INC-2 paper also outlines additional options for addressing awareness raising and education, information exchange, research, cooperation and coordination, and stakeholder engagement.

In INC-3, delegates proposed textual submissions to be included in a "revised zero draft," a mandate for the preparation of which was agreed upon.[3] INC-4 in April 2024 was expected to consider this revised draft, but they failed to settle on intersessional work to revise the draft ahead of INC-4.[4]

3.2.2 THE HIGH AMBITION COALITION TO END PLASTIC POLLUTION

A group of like-minded countries has taken the initiative to form a coalition of ambitious countries following the adoption of resolution 5/14, "End Plastic Pollution: Towards an International Legally Binding Instrument (ILBI)," by the UN Environment Assembly in March 2022.[5]

Norway and Rwanda co-chair the High Ambition Coalition. The coalition now has 60 members: Rwanda, Norway, Canada, Peru, Germany, Senegal, Georgia, Republic of Korea, the UK, Switzerland, Portugal, Chile, Denmark, Finland, Sweden, Costa Rica, Iceland, Ecuador, France, the Dominican Republic, Uruguay, Ghana, Monaco, Slovenia, the United Arab Emirates, Republic of Ireland, Seychelles, the

Netherlands, Belgium, Luxembourg, Cabo Verde, Burkina Faso, Australia, Azerbaijan, Colombia, Austria, Greenland, Jordan, Panama, Mali, New Zealand, Bulgaria, Montenegro, Cook Islands, Mexico, Guinea, Antigua and Barbuda, Armenia, Maldives, Federated States of Micronesia, Nigeria, Romania, Gabon, Japan, Mauritius, Spain, Estonia, Palau, Israel, and the European Union.[6]

The High Ambition Coalition to End Plastic Pollution is committed to developing an ambitious ILBI based on a circular approach along the entire life cycle of plastics. Their ambition is to end plastic pollution by 2040. At the outset of the plastic treaty negotiations, they have outlined three strategic goals and deliverables for success. The three strategic goals are:

1. **Restrain** plastic consumption and production to sustainable levels, including control measures
2. Enable a **circular economy** for plastics that protect the environment and human health
3. Achieve environmentally sound **management and recycling** of plastic waste

The seven critical deliverables for success are:

• Eliminate problematic plastics, including by bans and restrictions
• Develop global sustainability criteria and standards for plastics
• Set global baselines and targets for sustainability throughout the life cycle of plastics
• Ensure transparency in the value chain of plastics, including for material and chemical composition
• Establish mechanisms for strengthening commitments, targets, and controls over time
• Implement monitoring and reporting at each stage through the life cycle of plastics
• Facilitate practical technical and financial assistance and scientific and socioeconomic assessments

3.2.3 Oceanic Society

The Oceanic Society works to improve ocean health by deepening the connections between people and nature to address the root cause of its decline: human behavior.[7] The nonprofit organization provides seven plastic pollution solutions that everyone can take part in:

1. Reduce your use of single-use plastics
2. Support legislation to curb plastic production and waste
3. Recycle properly
4. Participate in (or organize) a beach or river clean-up
5. Avoid products containing microbeads
6. Spread the word to increase awareness
7. Support organizations addressing plastic pollution

3.2.4 TNO: FROM PLASTIC-FREE TO FUTURE-PROOF PLASTICS

The growing demand for plastics necessitates a structural shift.[8] To achieve this, TNO (Dutch Organization for Applied Scientific Research) has written a white paper *From #plasticfree to future-proof plastics | How to use plastics in a circular economy* in collaboration with Fraunhofer UMSICHT (German Institute for Environmental, Safety, and Energy Technology) describing four strategic approaches for a circular economy. Additionally, they have launched the "European Circular Plastics Platform" to address barriers and exchange solutions. The main goal of their strategy is a resilient, circular, and sustainable plastics economy.

The four strategic approaches for transforming the current mainly linear plastics economy into a circular economy are (Figure 3.2 and Figure 3.3):

- Narrowing the loop, as a first step, refers to reducing the amount of materials mobilized in a circular economy
- Operating the loop refers to using renewable energy, minimizing material losses, and sourcing raw materials sustainably
- Slowing the loop, by which measures are needed to extend the useful lifetime of materials and products
- Closing the loop, for which plastics must be collected, sorted, and recycled to high standards

The structural shift can only succeed if science, industry, politics, and citizens work together.

3.2.5 UNEP *TURNING OFF THE TAP* SYSTEMS CHANGE

This United Nations Environment Programme (UNEP) report proposes a system change scenario needed to address the causes and impacts of plastic pollution (Figure 3.4 and see Figures 2.17 and 2.18).[9]

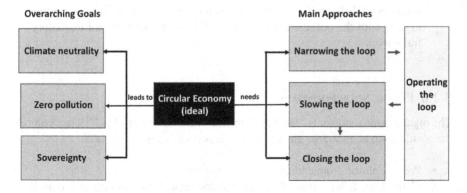

FIGURE 3.2 An ideal circular economy will lead to the achievement of the three overarching goals while incorporating the four main approaches along the loop (TNO, 2023).

Well-Known Approaches   **New Approaches**

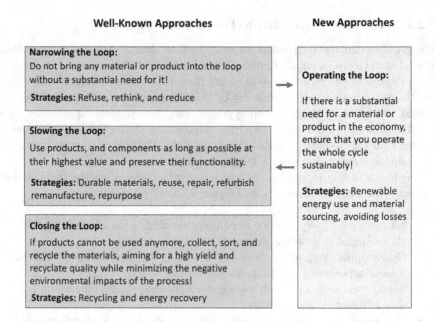

FIGURE 3.3 Detailed description of the three well-known approaches and the new approach for a sustainable circular economy (TNO, 2023).

The proposal includes:

- Reducing problematic and unnecessary plastic use
- Redesigning the system, products, and their packaging
- Transforming the market toward circularity in plastics

This can be achieved by accelerating three market shifts:

- Reuse (toward a reuse society from a throwaway economy)
- Recycle (toward recyclable products)
- Reorient and diversify (toward sustainable alternatives) and take actions to deal with the pollution legacy, all within sound policies and regulatory framework

3.2.6 PLASTIC SOUP FOUNDATION

The organization's slogan is "No Plastic in Our Water or Our Bodies."[10] They combat the plastic soup at its source and therefore focus on three main goals:

- Prevent plastic from ending up in the environment
- Share knowledge about the health risks related to plastics and plastic additives
- Achieve absolute reduction in the production and use of plastic

FIGURE 3.4 UNEP *Turning Off the Tap* report (UNEP, 2023).

3.2.7 Analysis and Synthesis of the Proposed Goals

The structural solutions proposed by four key players to combat plastic pollution include:

- Designing for circularity and strengthening waste management by the INC
- Reducing production and operating three market shifts toward circularity—reorient and diversify, reuse, and recycle—by the UNEP
- Reducing production and adopting four approaches toward a circular economy: narrowing the loop, operating the loop, slowing the loop, and closing the loop by TNO/Fraunhofer
- Reducing production, enabling a circular economy, and strengthening waste management by The High Ambition Coalition

Standard solutions between these groups are:

- Transition toward a circular plastic economy (reuse, recycle, and redesign), including sustainable waste management
- Actions to deal with existing pollution
- Sound regulatory framework

However, targets for reducing plastic production and use remain the object of debate among members of INC-2.

Conclusively, the following objectives and measures, which can be grouped into three categories along their life cycle (Figure 3.5), are proposed worldwide[11]:

1. **Upstream solutions**
   - Reducing plastic production and use, particularly problematic and unnecessary plastic products
   - Preventing and reducing plastic pollution at its origin, including primary microplastics such as microbeads
   - Awareness-raising initiatives and education

2. **Midstream solutions:**
   - Developing sustainable alternatives in materials
   - Creating incentives for the private sector to move from linearity to circularity
   - Enhancing innovation, transfer, and dissemination of environmentally sound technologies and best practices
   - Conducting thorough social and environmental impact assessments
   - Eliminating the release and emission of plastics into water, soil, and air
   - Promoting the extended producer responsibility

3. **Downstream solutions:**
   - Strengthening waste management
   - Creating new value chains for collecting, sorting, reusing, and recycling plastic
   - Addressing existing plastic pollution
   - Protecting human health

FIGURE 3.5 Key elements of a life cycle approach to addressing plastics pollution (Life Cycle Initiative, 2023).

In addition, goals in terms of timetables, roadmaps, and practical measures include:

- Operating the ILBI under a sound regulatory framework
- Finding an agreement on the ILBI by 2024 while defining how the instrument would build a legal mandate, specifying implementation measures such as national action plans, national reporting, and control measures along the entire life cycle
- Ending plastic pollution by 2040[12]

French philosopher Roland Barthes wrote, "More than a substance, plastic is the idea of its infinite transformation." It should now transform into sustainability.

3.3 RECYCLING IN THE CONTEXT OF A CIRCULAR ECONOMY

3.3.1 OVERVIEW

Efficient recycling of plastic waste supports the transition toward the circular economy while contributing to reducing greenhouse gas (GHG) emissions.[13] Recycling plays a significant role in today's standards of welfare. It saves energy and natural resources, including petroleum, land, plants, minerals, water, and air.

Collection and sorting are the first steps in ensuring that separated plastic products are delivered to recycling plants. Improvements in collection schemes and sorting technologies are essential to achieve higher recycling rates. Recycling plastics reduces the need to use fossil fuels to make new plastics, and using recycled materials to create new products is one of the best ways to decrease the environmental impacts of products.[14] Recycling PET and high-density polyethylene (HDPE) plastics can save 75–88% of the energy used to make virgin plastics and reduce GHG emissions by 70%. However, only 9% of global plastic waste is being recycled. In comparison, 50% is landfilled, 22% is mismanaged, 19% is incinerated, and only 6% of plastics are of recyclable origin today (UNEP, 2023, Turning Off the Tap; OECD).

Currently, analyses estimate losses in recycling processes around 25%; improved technologies can significantly reduce them. Despite increased global efforts to tackle the plastic waste crisis, the scale-up of recycling technologies is still needed to address plastic pollution and decarbonize the supply chain.

The Alliance of Mission-Based Recyclers (AMBR), a coalition founded by four of the original pioneers of mission-driven, community-based nonprofit recycling in the United States, wants to be a bridge from a throw-away society to a circular economy. They believe we cannot simply recycle out way out of the plastics pollution crisis. Their platform for improving recycling and reducing plastic consumption includes[15]:

- Eliminate problematic and unnecessary plastics from production
- Promote truth in labeling: remove the chasing arrows from nonrecyclable plastics
- Provide convenient recycling to all residents and businesses
- Improve plastics recycling facilities and processes
- Scale up reuse and refill solutions
- Refute false solutions such as plastics-to-fuel technologies
- Innovate new circular solutions in packaging, reuse, and recycling

The solution is to look "upstream" at how products can be redesigned to be more resource-efficient, made from recycled materials, and be more accessible to recover and to look "downstream" at how systems and infrastructure can better recover and remanufacture materials into new products.

Chemical Recycling Europe (CRE) is a platform that encourages its members to unite around the common goal of closing the loop for the plastics industry through technological innovation and participation in chemical recycling. The platform promotes the goal of recycling all plastic waste into its original components or value-added materials.[16]

3.3.2 TECHNOLOGIES

Plastic recycling technologies can be classified into mechanical, biochemical, chemical, thermochemical, carbon capture and utilization (CCU), and other advanced recycling technologies (Figure 3.6).[17]

Advanced recycling solutions cover technologies such as extrusion, dissolution, solvolysis, enzymolysis, pyrolysis, thermal depolymerization, gasification, and incineration with CCU.[18]

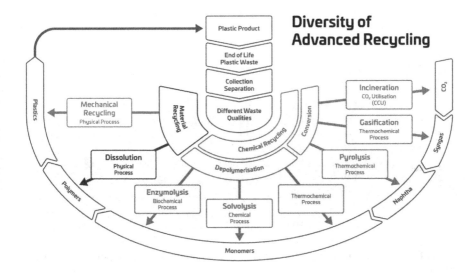

FIGURE 3.6 Diversity of recycling technologies (nova-Institute, 2023, with permission).

3.3.2.1 Mechanical Recycling

Mechanical recycling is the most common approach to plastics recycling. It is an essential tool in an environmentally and economically sustainable plastics economy, but current mechanical recycling processes are limited by cost, degradation of mechanical properties, and inconsistent product quality.[19] During the process, plastic waste is sorted, washed, shredded, heated, extruded, and pelletized with a chain degradation depending on the polymer and extrusion conditions. Extrusion is the foremost method used in mechanical recycling industries to produce regranulated material. It is cheap, large-scale, solvent-free, and applicable in principle to all types of thermoplastics. The primary degradation mechanism is the formation of radicals along the polymer chain due to oxygen-induced peroxy radicals and thermally induced abstraction of hydrogen atoms (Figure 3.7).[20]

The main limitations of mechanical recycling are that it cannot optimally recycle the variety of mixed plastic waste streams and does not efficiently control contamination from different plastic sources. At present, these limitations partly explain the low recycling rate of plastics.

3.3.2.2 Chemical Recycling

Chemical recycling is a growing alternative to mechanical recycling that implements circularity for plastics (Advanced Recycling Conference, 2023). It typically uses chemical depolymerization processes. Long hydrocarbon chains are broken down into shorter hydrocarbon fractions such as monomers by chemical processes, including, in the broadest sense, thermochemical, enzymatic, and solvolysis processes. These shorter molecules can be used as feedstock for new chemical reactions to produce new recycled plastics and other chemicals. The technology suits products that cannot be recycled mechanically, including mixed polymers and/or contaminated plastics.

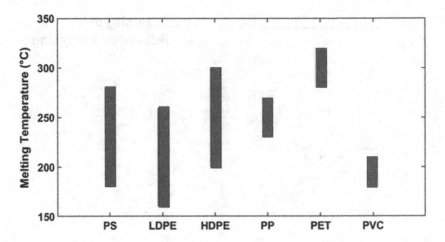

FIGURE 3.7 Processing ranges for the six most common packaging polymers (Schyns, 2020).

CEFIC, the European Chemical Industry Council, has developed a five-point plan that needs to happen for chemical recycling to contribute to the EU's circular economy[21]:

1. Create an EU single market for waste and end-of-life resources
2. Recognize all recycling technologies across all relevant EU legislation
3. Recognize a verified mass balance approach to calculate chemically recycled content in plastics and chemicals
4. Ensure the reliability and workability of mass balance
5. Drive investments into plastic recycling R&D programs and new business models.

3.3.2.3 Thermochemical Recycling

Thermochemical recycling uses heat to break down polymer chains through pyrolysis, gasification, and hydrocracking. It offers an alternative to conventional methods of treating plastic waste, such as mechanical recycling.[22] This process allows theoretically unlimited recycling of any plastic material (sorted or mixed), does not require extensive sorting practices, and operates under flexible conditions with low direct environmental impacts. However, thermochemical recycling is an energy-intensive process requiring electricity or fuel.

Pyrolysis is a core technology used in thermochemical recycling (Milton Roy). It involves decomposing plastic waste through heat, typically at around 500 °C without oxygen. This process requires conditioning and vaporizing plastic waste in the pyrolysis reactor. It is then condensed to produce mainly pyrolysis oil and fuel gas, char (carbon-black), and hydrocarbons. Pyrolysis produces less waste than traditional recycling methods while generating reusable products. Pyrolysis oil is a feedstock to make new plastics and synthetic chemicals.

In gasification, plastic waste reacts with gasifying agents (e.g., steam, oxygen, and air) at high temperatures around 500 – 1300 °C, producing synthetic gas, or syngas.[23] Syngas can be further processed to create new recycled plastics, fuels, and chemicals.

3.3.2.4 Enzymatic Recycling

In the coming years, enzymatic recycling will play a crucial role in solving the challenges faced by industries using difficult-to-recycle materials, which urgently need diverse and innovative solutions to transform these traditionally linear waste streams into circular ones.[24] This sector has recently seen a significant increase in transactions and investment, which indicates the urgency to address growing volumes of plastic waste and the increased demand for high-quality recycled plastic polymers and resins.

Recently, the U.S. National Renewable Energy Laboratory (NREL) discovered variants of enzymes adapted to deconstructing all varieties of PET, even durable crystalline varieties.[25] The laboratory discovered enzymes that could make it cheaper to recycle waste PET textiles and bottles than making them from petroleum. The enzymatic degradation of PET is being studied intensively because of the ability of hydrolase enzymes to depolymerize PET into its constituent monomers near the polymer's glass transition temperature. The research expands the number and diversity of thermotolerant hydrolase scaffolds for enzymatic PET degradation (Figure 3.8).

Carbios, a French biochemical company, pioneered designing and developing enzymatic processes to rethink the end-of-life of plastics and textiles in a circular economy. The company has the ambition to become the world leader in PET recycling.[26] The Carbios enzymatic recycling process uses an enzyme capable of specifically depolymerizing the PET present in various plastics and textiles.[27] Current thermomechanical recycling processes have limitations: only clear plastic can be recycled, with a loss of quality in each cycle, making it difficult to use 100% recycled PET for new products. Unlike conventional processes, the new process allows the recycling of all types of PET waste and the production of 100% recycled and 100% recyclable PET products without losing quality. The monomers resulting from the depolymerization are purified to be repolymerized into a PET of a quality equivalent to virgin fossil-based PET.[28]

In February 2022, Carbios and Indorama Ventures, the world's largest PET producer, jointly announced a collaboration to build a production plant operating Carbios' PET bio-recycling technology at Indorama Ventures' PET production site in Longlaville, France.[29]

The plant will have a capacity of, tonnes of PET waste per year, equivalent to 2 billion bottles, and is expected to be operational in 2025. An exclusive, long-term partnership with Novozymes, the world leader in enzyme production, was announced in January 2023.[30] In November 2023, Carbios obtained the building and operating permits for the plant.[31]

3.3.2.5 Dissolution Recycling

Dissolution recycling is a solvent-based technology in which the polymer present in a mixed, multimaterial plastic waste is selectively dissolved in a solvent, allowing it to be separated from the waste and recovered in a pure form without changing its chemical nature (Plastics Europe, 2023). Several industrial examples of this

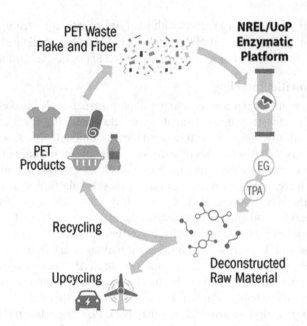

FIGURE 3.8 Enzymatic platform that deconstructs waste PET back into its building blocks, terephthalic acid (TPA) and ethylene glycol (EG), as developed by National Renewable Energy Laboratory (NREL) and University of Portsmouth (UoP). (Figure by Elizabeth Stone, NREL, 2022.)

technology already exist and apply to different polymers, such as polyvinyl chloride (PVC), polystyrene (PS), nylon (PA,) or polypropylene (PP).

3.3.2.6 Organic Recycling: Composting and Anaerobic Digestion

Organic recycling refers to the microbiological treatment of organic waste, including biodegradable plastic waste, under aerobic conditions (composting) or anaerobic conditions (anaerobic digestion) (Plastics Europe, 2023).[32]

Industrial composting is carried out under controlled conditions. The composting process is governed by several factors, including temperature (typically 50–60 °C), moisture, amount of oxygen level, and particle size.[33] The conditions in industrial composting differ from those of home composting, in which, for example, the temperature tends to be lower. The outputs of the industrial composting process are CO_2, water, and compost. The compost contains nutrients and can be used, for example, in agriculture to improve soil quality. The benefits of industrial composting include the absence of chemicals in the process, the replacement of mineral fertilizers, and the sequestration of carbon in the soil.

An industrially compostable material must comply with EN 13432, which sets out the requirements for bioplastics to be certified as compostable (see Section 3.4.1 "Biodegradability and Compostability"). ISO 14855 tests the ultimate aerobic biodegradability of plastic materials under controlled composting conditions by analyzing the carbon dioxide generated.

Anaerobic digestion applies to organic waste and specific polymers that microorganisms can convert into biogas, a mixture of mainly methane and carbon dioxide (Figure 3.9).

3.3.2.7 Solvolysis

Solvolysis, as applied to plastic waste, is another new recycling technology. It includes depolymerization processes such as alcoholysis, hydrolysis, acidolysis, aminolysis, and various exchange reactions that produce oligomers or monomers.[34] Solvolytic techniques fall into the category of chemical recycling. Suitable candidates are mostly step-growth thermoplastics and thermosets: polyesters, polyamides, and polyurethanes.

3.3.2.8 Carbon Capture and Utilization

Carbon capture technologies have been developed to capture carbon emissions at their source.[35] CCU is one essential pillar of renewable carbon supply alongside

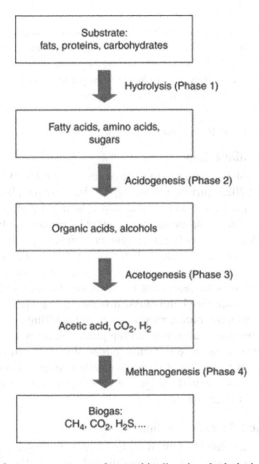

FIGURE 3.9 The four process stages of anaerobic digestion: hydrolysis, acidogenesis, acetogenesis, and methanogenesis. Different microorganisms are responsible for the processes in each stage.

biomass and recycling. It is, therefore, one of the critical technologies for the transition to sustainable chemical and fuel production, replacing fossil feedstocks and combating climate change.[36] The CCU industry proposes a solution to both rising CO_2 emissions and increasing plastic consumption: creating lower-carbon, degradable polymers using CO_2 emissions as the feedstock.[37] There are at least three major pathways to convert CO_2 into polymers: electrochemistry, biological conversion, and thermocatalysis (Figure 3.10).

The latter is the most mature CO_2 utilization technology, where CO_2 can be used either directly to yield CO_2-based polymers, in particular, biodegradable linear-chain polycarbonates, or indirectly, through the production of chemical precursors (such as methanol, ethanol, acrylate derivatives, or mono-ethylene glycol) for polymerization reactions.

According to the Nova Institute, more than 1.3 million tonnes of global capacity for CO_2-based products existed in 2022, and this is expected to at least quadruple by 2030. The institute has drawn a graph showing how CO_2 can be used for chemicals and polymers (Figure 3.11). The conversion technologies shown are chemical catalysis, Fischer-Tropsch, hydrogenation, biochemistry, and electrochemistry.

Many industrialists are moving to the idea of gathering and reusing CO_2 in the air to produce new materials. However, the task is far from easy, and some large companies abandon their CO_2 recycling projects to move to other technologies, such as chemical dissolution or biobased materials.

3.3.3 Acceleration of Recycling

3.3.3.1 Design Optimization

For many years, the slogan has been to increase recycling rates and eliminate single-use plastics (UNEP, 2023, Turning Off the Tap p. 25). However, plastic products must be designed and made from materials that allow them to be recycled. Nearly 80% of the plastic in short-lived plastic products cannot be economically recycled due to design choices such as additives (e.g., dyes) and material composition. A tiny fraction of plastic products can be safely reused. Establishing design rules, e.g., to reduce the multipolymer plastics, to favor the design formats that are easier to reuse or recycle, or to standardize formats for reuse, can improve the economics of reuse and recycling schemes and reduce GHG emissions. It is shown that GHG emissions can be lowered by ~48% when comparing recycling versus landfilling.

There needs to be more than just designing plastic products to be recyclable; collection systems must be in place to facilitate recycling. It is estimated that around 2 billion people are not connected to waste collection systems. Coordinating collection and sorting processes with the recycling system can ensure that recycled plastic is the same quality as virgin plastic.

3.3.3.2 Economics: Basel Convention

Recycling can be accelerated by improving its economics, aligning design incentives with the recycling economy, and ensuring safe and fair recycling in practice and at scale. As with reuse, certain chemicals of concern in plastics reduce their potential for circularity and thus make recycling less economically viable. The trade of plastic waste from areas without recycling infrastructure to places with surplus recycling

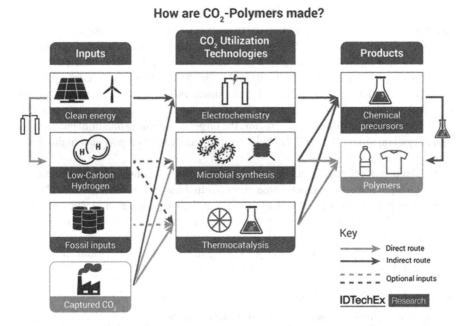

FIGURE 3.10 Pathways to polymers from CO_2 (IDTechEx, 2023).

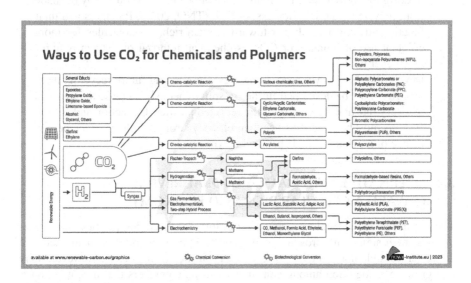

FIGURE 3.11 Ways to use CO_2 for chemicals and polymers (Nova-Institute, 2023, with permission).

capacity can increase circularity. By establishing a legally binding framework for the trade in plastic waste, the Basel Convention plastic waste amendments create the conditions for more transparent global trade in plastic waste.[38] The amendments also strongly incentivize the private sector, governments, and other stakeholders to develop an environment that enables recycling and reduces plastic waste generation.

3.3.3.3 Ellen MacArthur Foundation

There are opportunities to accelerate the transition to a circular economy, particularly recycling. For example, over 1000 organizations have signed up to the Ellen MacArthur Foundation's New Plastics Economy vision, pledging to make their packaging products reusable, recyclable, or compostable by 2025.[39] However, according to the 2022 New Plastics Economy Global Commitment progress report, the commitment will likely not be met.[40] The use of flexible packaging and a lack of investment in collection and recycling infrastructure means the target is becoming unattainable for most signatory companies. The 2022 Global Commitment progress report also highlights:

- The use of recycled content in plastic packaging continues to rise strongly, having doubled in the past three years, from 4.8% in 2018 to 10% in 2021.
- Over half of business signatories have cut their use of virgin plastics since 2018, but overall use among the group increased in 2021 back to 2018 levels.
- The share of reusable plastic packaging decreased slightly to an average of 1.2%.

3.3.3.4 Future R&D Actions

Future R&D activities toward a circular economy include:

- Redesigning polymers: In a circular economy, plastics should only be made from recycled material, biomass, or CO_2 (TNO, 2023). Polymers and their functionality must be driven toward oxygen-rich(er) molecules for more efficient use of biomass or CO_2, as chemical removal of oxygen requires much energy.
- Developing new recycling technologies: Current and future recycling technologies must be developed further to simultaneously increase the material recovery rate and the quality of the recycled plastics. Collection and sorting systems must be included since their efficacy strongly influences the performance of mechanical and chemical recycling of plastics.

3.4 ALTERNATIVE MATERIALS: A GLOBAL REVIEW TOWARD A CIRCULAR BIOBASED ECONOMY

The ocean has increasingly become a repository for discarded plastics and microplastics, with significant proven social, economic, and environmental impacts.[41] It is neither possible nor desirable to remove all plastics from society. However, there is a growing recognition among consumers and policymakers that urgent action is needed to stop the flow of single-use plastics and that alternatives can play a significant role in reducing our dependency. However, awareness of plastic pollution is growing everywhere, and options for plastics, especially single-use plastics, should be intensively explored.

Increased use of biomass resources to produce plastics and chemicals effectively reduces global warming and the depletion of fossil resources. However, the impact of such increased use of biomass on plastic pollution is more controversial. Biobased plastics refer to plastics containing materials wholly or partly of biogenic origin, while bioplastics refer to biobased and biodegradable plastics.

3.4.1 BIODEGRADABILITY AND COMPOSTABILITY

3.4.1.1 Definition

Biodegradability is defined as the ability of a material to break down after interaction with microorganisms and enzymes.[42] Although compostability is a term used for products that can be processed in an industrial composting plant, it is sometimes used to refer to home composting and even to biodegradation in the natural environment (Sulapac, 2023). One way to recycle biodegradable materials is through organic recycling, which includes industrial composting and anaerobic digestion.

Both compostable plastics and biodegradable plastics are materials that break down into their organic components.[43] However, composting some compostable plastics requires strict control of environmental factors such as those found in industrial composting facilities. Biodegradable materials can take an infinite amount of time to break down, while compostable materials will decompose into natural elements within a specific time frame.[44]

Generally, the adherence of microorganisms on the surface of plastics, followed by the colonization of the exposed surface, are the central mechanisms involved in the microbial degradation of plastics.[45] The enzymatic degradation of plastics by hydrolysis is a two-step process: first, the enzyme binds to the polymer substrate, and subsequently, the enzyme catalyzes a hydrolytic cleavage. Polymers are degraded into oligomers and monomers and mineralized to CO_2 and H_2O.

Both the chemical and physical properties of plastics influence their biodegradation. The surface conditions (surface area, hydrophilic, and hydrophobic properties), the chemical structures (molecular weight and molecular weight distribution), and the physical structures (glass transition temperature, melting temperature, elastic modulus, crystallinity, and crystal structure) of polymers play an essential role in the biodegradation processes. The higher the molecular weight, the higher the crystallinity; the higher the melting temperature, the higher the branching and the lower the biodegradability.

3.4.1.2 Oxo-biodegradability

Oxo-degradable plastics, also known as oxo-biodegradable plastics, are made from conventional plastics and supplemented with specific additives to mimic biodegradation.[46] Typical oxo-degradable additives include organic salts of transition metals such as iron, nickel, cobalt, or manganese, which react with oxygen to promote the breaking of chemical bonds in the molecular structure of plastic.[47] However, these additives only facilitate fragmentation of the materials, which do not fully degrade but break down into very small fragments that remain in the environment. This process would be more accurately described by the term "oxo-fragmentation." Oxo-degradable plastics contribute to microplastic pollution, which poses an environmental risk, particularly in the oceans.[48] These products cannot be marketed as compostable and biodegradable, and they are often more harmful to the environment than traditional plastic.

3.4.1.3 Testing

Biodegradation testing quantifies the biochemical degradation of the material as microorganisms consume it in a specific environment, the composting facility.[49] The test indicates the biochemical conversion of solid or liquid organic carbon to gaseous CO_2. The test is performed according to the ISO 14855 (EN ISO 14855) standard, "Determination of the ultimate aerobic biodegradability of plastic materials under controlled composting conditions. Method by analysis of evolved carbon dioxide." This method is designed to simulate typical aerobic composting conditions and provide the percentage conversion of the carbon in the test material to evolved carbon dioxide and the conversion rate.[50]

An industrially compostable material (packaging) must comply with EN 13432 (EN ISO 13432) "Packaging—Requirements for packaging recoverable through composting and biodegradation—Test scheme and evaluation criteria for the final acceptance of packaging," which sets out the requirements for bioplastics (packaging but also other applications) to be certified as compostable. According to EN 13432, compostable plastics disintegrate (i.e., fragmentation and loss of visibility; residual mass less than 10% of the original mass) after 12 weeks and biodegrade by at least 90% after six months.[51] This means that 90% or more of the plastic material will have been converted to CO_2. The remainder is transformed into water and biomass—i.e., valuable compost with no negative impact on the composting process and within safe limits for heavy metals. The "Seedling" logo can recognize plastics certified according to EN 13432.[52]

3.4.1.4 Coordinate System for Bioplastics

Plastics can be classified according to a biodegradable/nonbiodegradable axis and a biobased/fossil-based axis (Figure 3.12).[53]

Some biobased polymers have the same structure as fossil-based polymers and are not biodegradable, such as:

- Polyethylene
- Polyethylene terephthalate
- Polyamide

While others have a new structure and are biodegradable, such as:

- Polylactic acid
- Polyhydroxyalkanoate
- Polybutylene succinate
- Plasticized starch (starch blends)
- Plasticized regenerated cellulose (cellophane)

Some polymers are fossil-based and biodegradable, such as polycaprolactone, aliphatic polyesters, and polyvinyl alcohol (PVA). Conventional plastics are fossil-based and not biodegradable.

In 2022, bioplastics accounted for less than 1% of the ~ 400 million tonnes of plastics produced annually. Global bioplastics production capacity is set to increase

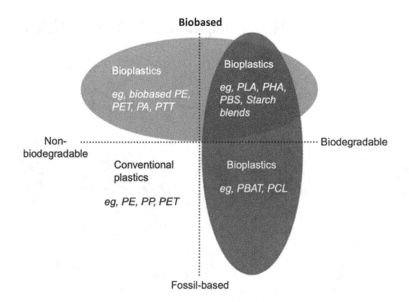

FIGURE 3.12 Classification of bioplastics by their biodegradability and origin. Definition of bioplastics. PE, polyethylene; PET, polyethylene terephthalate; PVC, polyvinyl chloride; PP, polypropylene; PTT, polytrimethylene terephthalate; PBAT, polybutylene adipate terephthalate; PCL, polycaprolactone; PBS, polybutylene succinate; PLA, polylactic acid; PHA, polyhydroxyalkanoate; PEF, polyethylene furanoate (European Bioplastics; with permission for previous book).

from around 2.2 million tonnes in 2022 to approximately 7.43 million tonnes in 2028 (Figure 3.13).[54] In 2022, biodegradable bioplastics accounted for 52% of the total capacity, and their relative share is predicted to increase further to 56% in 2027, mainly due to the development of PLA and PHA.

3.4.2 BIOBASED BUT NOT BIODEGRADABLE PLASTICS

The main non-biodegradable bioplastics are bio-PE, bio-PP, bio-PET, bio-polytrimethylene terephthalate (bio-PTT), and bio-PA.[55] They have the same chemical structure as their fossil-based counterparts. Like most conventional plastics, biobased plastics must be recycled separately for each material type (e.g., PET stream).[56] Where a recycling stream for a specific plastic type is established, the biobased alternatives can be recycled together with their conventional counterparts.

3.4.2.1 Polyethylene and Polypropylene

Bioethylene can be produced by the catalytic dehydration of bioethanol, produced by the fermentation of sugars (carbohydrates), and then polymerized to give bio-PE, as shown in Figure 3.14.[57]

Bio-propylene can be produced from bioethanol using a zeolite catalyst and then polymerized to give bio-PP.[58]

Global production capacities of bioplastics

in 1,000 tonnes

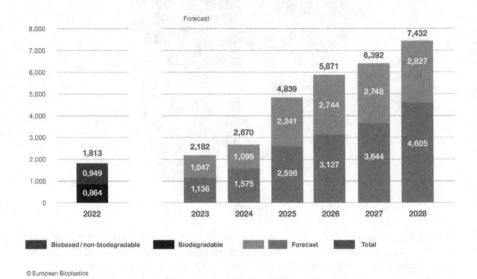

FIGURE 3.13 Worldwide production capacities of bioplastics in 2022 and forecast from 2023 to 2028 (European Bioplastics; Nova-Institute, 2023; www.european-bioplastics.org/market/, with permission).

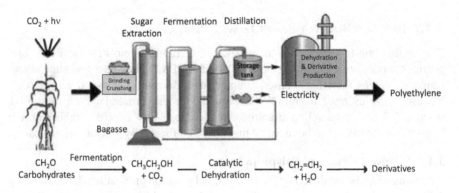

FIGURE 3.14 Schematic flow diagram of bio-polyethene production from sugarcane via fermentation into ethanol and subsequent dehydration into ethylene (IFT, 2014, with permission).

3.4.2.2 Polyethylene Terephthalate

Aromatic polyesters have excellent physical and mechanical properties compared to aliphatic polyesters, but their strong resistance to bacterial or fungal attack results in low degradability under environmental conditions. PET is produced from the poly-condensation of two monomers, 70% purified terephthalic acid (PTA) and 30% mon-oethylene glycol (MEG) (Figure 3.15).[59]

3.4.2.3 Polyethylene Furanoate

Polyethylene furanoate (PEF) is an aromatic polyester produced by polycondensation of 2,5-furandicarboxylic acid (FDCA) and ethylene glycol (Figure 3.16).[60]

PEF is a chemical analogue of PET, 100% recyclable, and 100% plant-based.[61] Although PEF is not biodegradable under industrial composting conditions, it

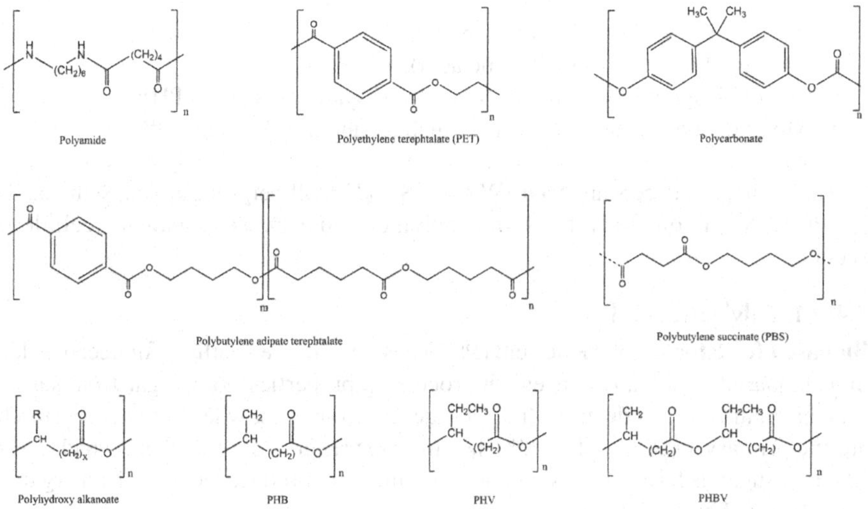

FIGURE 3.15 Polyamide, polyethylene terephthalate (PET), polycarbonate, polyadipate terephthalate, polybutylene succinate (PBS), polyhydroxyalkanoate, PHB, PHV, and PHBV.

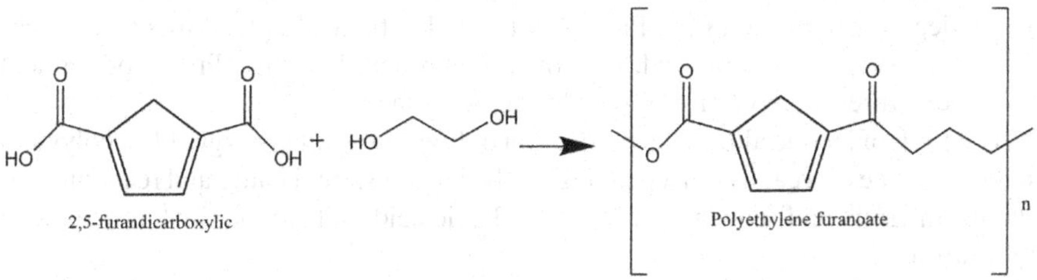

FIGURE 3.16 Polyethylene furanoate (PEF).

degrades significantly faster than PET, as described in the European standard EN 13432: under industrial conditions, 90% of PEF biodegraded within 240 and 385 days, in weathered and unweathered conditions, respectively.[62]

PEF was developed by the Dutch company Avantium. The world's first commercial FDCA plant is expected to be completed by 2024. Avantium uses its YXY Technology, which converts plant-based sugars into aromatic diacids. PEF can be employed in various applications, including bottles, packaging, films, fibers, and textiles.[63]

3.4.3 BIOBASED AND BIODEGRADABLE

Processes for the production of biobased and biodegradable thermoplastic polymers can be classified into three main groups based on the processes involved and also the types of the polymers:

- Chemical polymerization of monomers derived from biological processes, such as PLA, PGA (polyglycolic acid), and PBS
- Direct biosynthesis of polymers in microorganisms, such as PHA
- Modification of natural polymers, such as starch and cellulose[64,65]

The leading producers are NatureWorks US and TotalEnergies Corbion Netherlands for PLA, Novamont Italy for starch polymers, and Daicel Japan for cellulose acetate.

3.4.3.1 Polylactic Acid

Biobased feedstock. Polylactic acid (PLA) has been in the spotlight for decades due to its biobased nature and convenient processing properties. Raw sugar from sugarcane or sugar beet and dextrose (i.e., D-glucose) from corn, wheat, or cassava starch are the primary PLA feedstocks.[66] The sugar extracted from the plant, which fixes CO_2 into sugar, is fermented using microorganisms to produce lactic acid, an organic acid produced by the human body.

Synthesis. PLA is a thermoplastic polyester $(OCOR)_n$ with the formula shown in Figure 3.17, formally obtained by condensation of lactic acid CH3-CHOH-COOH with water loss.

However, the most common route to PLA is by ring-opening polymerization (ROP) of lactide, the cyclic ester dimer formed via lactic acid condensation, followed by the depolymerization of the PLA oligomer.[67] Lactic acid optical monomers consist of L-lactic and D-lactic acid, with the predominant L-form.[68] From both optical monomers, three stereo forms of lactide can be formed.

The ROP of lactide allows the production of high-molecular-weight PLA polymers in the presence of a catalyst. Depending on the lactide stereoisomer and reaction conditions, the ratio and sequence of D- and L-lactic acid units in the final polymer can be controlled.

Stereoregularity. Stereoregularity strongly affects the thermal and mechanical properties, as well as the crystallization behavior of PLA.[69-71] Isotactic PLA is

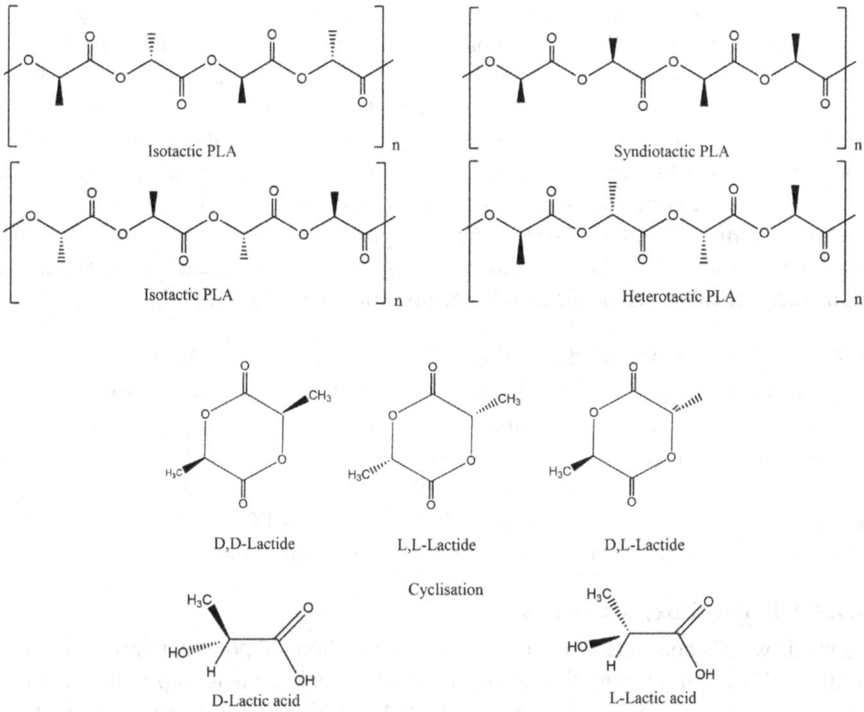

FIGURE 3.17 Polylactic acid with ester-bond backbone. The optical monomers of lactic acid along with the three stereo forms of lactide: (SS) L-lactide; (RR) D-lactide; (SR) meso-lactide.

formed from either pure D-lactide (leading to PDLA) or L-lactide (leading to PLLA), and the sequential stereocenters have the same absolute configuration.[72] Syndiotactic PLA has alternating configurations of the sequential stereocenters. Atactic PLA has a random distribution of configurations of the stereocenters. Still, its heterotactic counterpart (racemo diad [two units oriented in opposition] adjacent to a meso diad [two identically oriented units]) has regions of stereo-homogeneity.[73] An isotactic stereoblock PLA with blocks of opposite chirality, also called stereo-complexes (PLA-sc), is similar to isotactic PLA but differs in that racemic lactide is used instead of pure L-or D-lactide.[74]

Stereocomplexes. Several approaches have been developed to improve PLA thermal resistance: blends, nucleating agents, copolymers, and the most promising one, the formation of stereocomplexes. PLA-scs are formed by co-crystallizing the enantiomeric PLLA and PDLA chains, which are prerequisites for forming PLA-sc. The melting temperatures of the stereocrystals are 50 °C higher than those of the homocrystals, reaching 220–230 °C. The presence of hydrogen bonds contributes to the formation of the new co-crystalline structure. Future research should be carried out on various aspects of PLA-sc formation, including processing techniques.

Compostability. PLA is compostable but only in an industrial composting plant.[75] Under these conditions, PLA can be biologically degraded within a few days and up to a few months. Temperatures must be above 55–70 °C. In nature, it takes at least 80 years for PLA to decompose, which means that in the ocean and on land, it contributes not only to conventional petroleum-based plastics but also to environmental pollution from plastics and microplastics. For this reason, PLA should not be thrown into nature, home composters, or organic waste, just like other plastics.

Applications. PLA has a wide range of applications because its properties are similar to PS and PET. These include packaging as a significant use, agriculture, biomedical, textiles, electronics, and transportation. (Figure 3.18)[76,77].

3.4.3.1.1 Polyglycolic Acid and Polylactic Acid-Co-Glycolic Acid

PGA is a biodegradable, thermoplastic polymer and the most straightforward linear, aliphatic polyester. With a structure similar to PLA, PGA has promising properties, such as good biodegradability, and barrier properties but exhibits very different properties.[78] It can be produced from synthetic or natural glycolic acid (hydroxyacetic acid) using polycondensation or ROP. PLA with PGA can be modified via co-polymerization, physical blending, and multilayer lamination.

3.4.3.2 Polyhydroxyalkanoates

Polyhydroxyalkanoates (PHAs) are thermoplastic aliphatic polyesters produced naturally by bacterial fermentation of sugars and fatty acids. Bacteria produce them as energy storage molecules. The ability of PHAs to biodegrade in soil, compost, or water in a nontoxic manner has made them highly attractive for use as a replacement for synthetic polymers, particularly in packaging and other disposable products. The main bottleneck in the commercial viability of PHAs is their high cost compared to

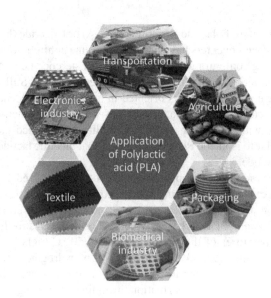

FIGURE 3.18 Applications of PLA (Ghomi, 2021).

their non-biodegradable synthetic counterparts. PHAs from waste feedstocks, rather than high-cost sugars and fatty acids, are promising for cheaper PHA and a circular economy (Figure 3.19).[79]

Defined by the carbon chain length of the 3-hydroxyalkanoate monomer, over 100 types of PHA monomers have been identified to date. PHAs can also exist as copolymers, which have superior properties to homopolymers. The most commonly used are poly(3-hydroxybutyrate) (PHB) and poly(3-hydroxyvalerate) (PHV) and their co-polymer poly(3-hydroxybutyrate-co-3-hydroxyvalerate) (PHBV) (Figure 3.18).

Synthesis. PHAs can be produced in three ways: fermentation of sugars and fatty acids by microbes, synthesis in genetically modified plants, and enzymatic catalysis. Fermentation is the most reliable of these options mainly because of relatively high PHA yields. Typical substrates for synthesizing PHAs include simple sugars and oils, such as glucose and glycerol.

Biodegradability. PHAs are considered the most readily biodegradable polymer among many biodegradable polymers such as PLA, PBS, and PBAT in aerobic environments (soil, compost, and water) and show promising biodegradable behavior in anaerobic (sewage sludge, digesters, and landfills) environments.[80] PHAs benefit from biotic degradation by several types of bacterial and fungal enzymes. However, PHAs are expensive and need much more research, including extraction from bacteria.

Applications. Some of the most interesting applications include single-use food packaging, single-use consumer goods packaging, biomedical, and agricultural (mulching).[81]

3.4.3.3 Polybutylene Succinate

Polybutylene succinate (PBS) is a thermoplastic aliphatic polyester with properties comparable to polypropylene (Figure 3.15).

Among aliphatic biodegradable polyesters, PBS is a promising material because it offers excellent biodegradability, melt processability, and chemical resistance.[82] However, PBS has limited applicability due to its thermal stability, high flammability, and relatively poor mechanical properties.[83]

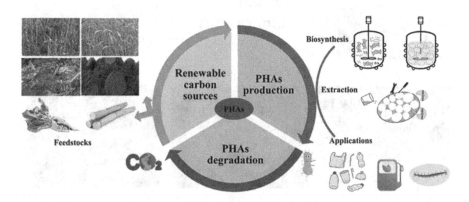

FIGURE 3.19 PHAs in a circular economy (Zhou, 2023).

Synthesis. Direct esterification of succinic acid with 1,4-butanediol is the most common way to produce PBS. Both monomers can be derived from biological or fossil-based sources. PBS consists of a two-step process. First, an excess of the diol is esterified with the diacid to form PBS oligomers with the elimination of water. The second step is using a catalyst for the polycondensation of oligomers to high-molecular-weight PBS.[84]

Biodegradability. The biodegradability of polyesters such as PBS is influenced by molecular weight, degree of crystallinity, and chemical structure (Aliotta, 2022). The hydrolysable ester bond in the main chain, susceptible to microbial attack, is the main reason for polymer biodegradation.

Applications. PBS, its copolymer with an adipic segment, and the blends with PLA, PHBV, or TPS have commercial applications in many fields, such as packaging, agriculture, fishing, forestry, construction, and electronics (Aliotta, 2022). Final applications include biomedical, food packaging, mulch film, and tableware (Figure 3.20).[85] PBS can also be used for monofilament, injection molding, tape, split yarn, and textile industries.

PBS composites filled with natural fibers will reduce the price of PBS products and have emerged as materials of interest in industries such as packaging, automotive, and construction.[86]

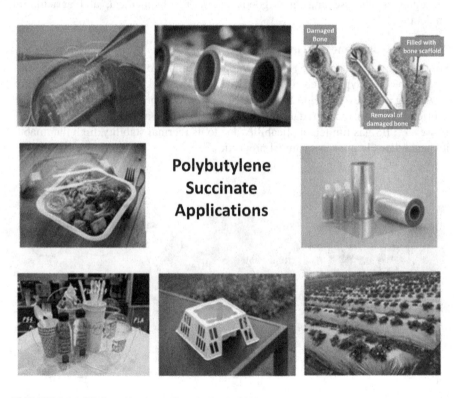

FIGURE 3.20 PBS applications (S.A. Rafiqah, 2021).

3.4.3.4 Starch

Starch, a biodegradable polymer that is abundant and inexpensive to produce, is one of the most promising candidates for bioplastics. Starch occurs naturally in semi-crystalline granules and is an energy store in plants. In its natural state, starch is not thermoplastic because it undergoes pyrolysis before reaching the melting point of its crystalline regions.

Starch is made up of two main components: amylose and amylopectin. Amylose comprises 15–35% of plant granules and is a predominantly linear polysaccharide with α-(1,4)–linked D-glucose units. Amylopectin is a highly branched molecule with α-(1,4)–linked D-glucose backbones and about 5% α-(1,6)–linked branches (Figure 3.21).[87]

Thermoplastic starch (TPS) is produced by mixing native starch with a plasticizer at a temperature above the starch gelatinization temperature, typically in the 70–90 °C range.[88] Heating native starch (50-100 °C) in excess water results in gelatinization. This hydrothermal process, in which water acts as a temporary plasticizer, changes the morphology of starch granules from an ordered to a disordered amorphous structure. Plasticizers come in various forms, such as glycerol, sorbitol, urea, fructose, sucrose, and glycol.[89] However, the most commonly used are from the polyol group, namely glycerol and sorbitol. Applications of TPS include biodegradable packaging, film material, and disposable tableware.

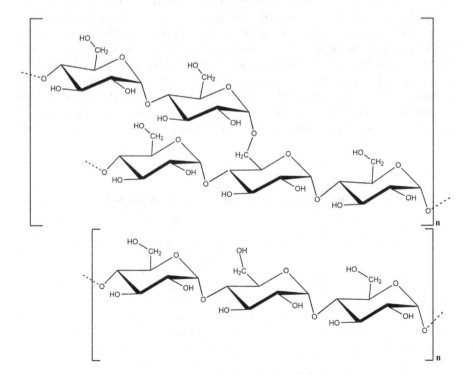

FIGURE 3.21 Schematic representations of amylose and amylopectin.

Blending starch with other biopolymers has been presented as a viable alternative to overcome the shortcomings of TPS, i.e., sensitivity to moisture and poor mechanical and barrier properties (Table 3.1). The degree of compatibility between starch and other biopolymers depends on the specific biopolymer. Blends of TPS and PLA offer significant advantages in terms of cost, properties, and biodegradability. The mechanical properties of the blends are often inversely related to their biodegradability. Incorporating natural fibers as fillers into the starch matrix is another promising option— fiber-reinforced TPS forms biocomposites. The high mechanical resistance of the fibers embedded in the starch matrix improves the mechanical properties of the biocomposite.

3.4.3.5 Cellulose

Cellulose is the world's most common biopolymer and is fully biodegradable and compostable. It will play a growing role as an alternative to fossil-based polymers and fuels. The biosphere is estimated to produce around 100 billion tons of cellulose annually. Cellulose is the structural component of plant cell walls, where it occurs in the form of crystalline microfibrils a few nanometers wide.[90]

Its primary structure is a homopolymer consisting of long linear chains of β-(1-4)–linked D-glucose residues associated with the repeating disaccharide unit of cellobiose (Figure 3.22). Cellulose in its native state is not thermoplastic because it does not soften or melt upon heating. Instead, it decomposes at high temperatures.

Whereas native cellulose, known as cellulose I polymorph, generally occurs as fibers, regenerated cellulose (cellulose II polymorph), obtained by its dissolution and precipitation, is manufactured as fibers, films, or other products with a morphology differing from native cellulose. The morphology of these regenerated products depends on the solvent used, the regeneration process, and the subsequent product processing. Mercerization, an alkaline treatment of cellulose fibers, also converts cellulose I into cellulose II, keeping the fundamental fibrillar morphology.

Since cellulose cannot melt and is not thermoplastic, the dissolution of cellulose is essential for producing fibers, films, sponges, casings, tire cords, etc. Rayon, lyocell, and cellophane are examples of regenerated cellulose products.

TABLE 3.1

Applications of Starch Blends and Composites

Blends	Application
Starch/PVA	Water-soluble laundry bags
	Biomedical and clinical field
	Replacement of polystyrene
Starch/PLA	Biodegradable tray
	Electronic devices
	Pharmaceutical
Starch/PBS	Packaging material
	Fishery
	Automotive
Starch/natural fibers	Food packaging
	Biodegradable material

Source: Diyana, 2021.

The solvents for cellulose can be classified into two categories: non-derivatizing and derivatizing systems. Among non-derivatizing solvents are:

- The N-methyl morpholine-N-oxide (NMMO)/water system, which has led to the lyocell process and products
- Ionic liquids
- Aqueous sodium hydroxide at low temperatures
- N, N-dimethylacetamide/lithium chloride
- Cuprammonium, giving rise to cuprammonium rayon, invented in 1890[91–93]

Among the derivatizing solvents is the industrial carbon disulfide/aqueous NaOH system, which has led to the viscose process and its multiple products, including rayon fibers and cellophane films commonly used for food packaging.

Cellulose is a chemically stable polymer.[94] However, it can be degraded in nature by enzymes as an essential part of the carbon cycle. In a typical cellulose-degrading ecosystem, cellulolytic bacteria and fungi convert insoluble cellulose to soluble sugars, primarily cellobiose and glucose, which are assimilated by the cell. To catalyze this process, the cellulolytic microorganisms produce a wide variety of enzymes known as cellulases. Cellulases catalyze the hydrolysis of the β-1,4-glucosidic linkages. The commercial potential for efficient enzymatic hydrolysis is enormous, particularly in using cellulosic biomass as a renewable energy source.

Certificates show that all cellulose-based fibers rapidly biodegrade in all test environments (soil, industrial composting, home composting, fresh water, and marine water) within the time periods set by the relevant standards. Cellulose-based fibers offer an alternative to combat plastic pollution. Cellophane meets the European EN 13432 and American ASTM D6400 composting norms.[95] Compostability certification ensures that the products will completely degrade within 90 days in a composting facility.

Chemical modification is needed to impart thermoplastic properties to cellulose. It also affects the biodegradability of natural polymers. Chemical modifications of

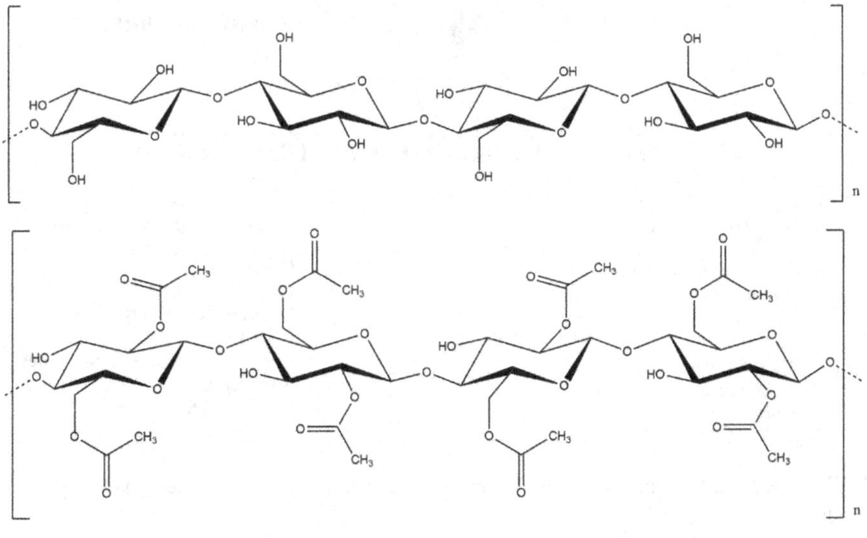

FIGURE 3.22 Schematic representation of one cellulose chain and one cellulose acetate chain.

cellulose involve reaction with its hydroxyl groups, which undergo esterification and etherification. Cellulose esters are soluble in a wide range of solvents.

Cellulose acetate (CA) is an essential cellulose ester due to its applications in fibers (cigarette filters and textile fibers), films, packaging, coatings, and membranes. It is generally obtained by reacting and dissolving pulp with acetic anhydride, using acetic acid as a solvent and sulfuric acid as a catalyst. Cellulose diacetate and cellulose triacetate are mistakenly referred to as the same fiber; although they are similar, their chemical identities differ. Triacetate is known as the generic or primary acetate containing no hydroxyl group. Acetate fiber is known as modified or secondary acetate having two or more hydroxyl groups. The high degree of acetylation is crucial for good thermoplastic properties. At the same time, a high degree of acetylation lowers the degradation potential. CA requires the presence of esterases for the first step in biodegradation (Figure 3.23).[96]

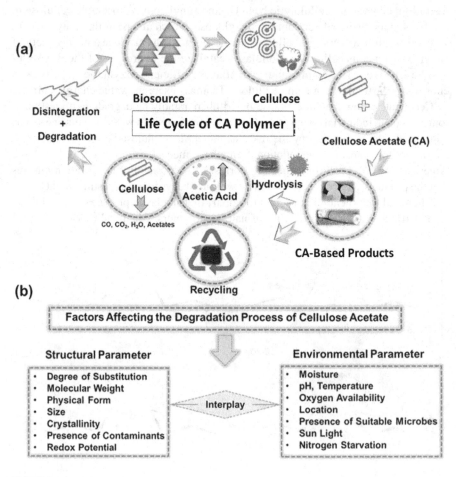

FIGURE 3.23 Life cycle of cellulose acetate and parameters of the degradation process (Yadav, 2021).

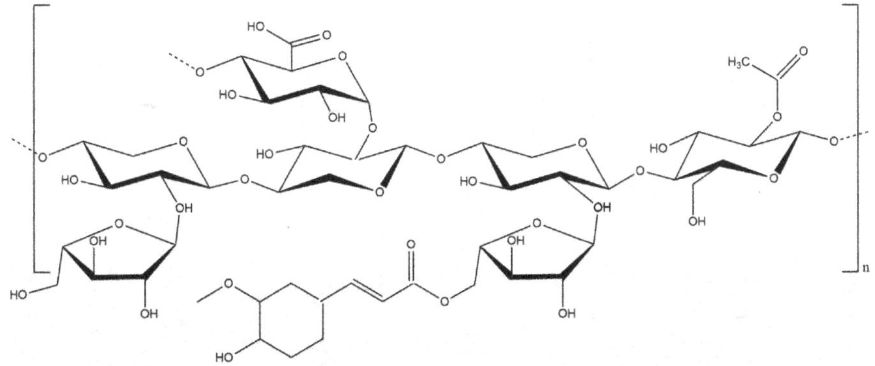

FIGURE 3.24 General structure of xylans, which are decorated at O2 with 4-O-methyl-d-glu-curonic acid (4-O-MeGlcA) and at O2 and/or O3 with arabinofuranose (Araf), whereas the backbone can also be acetylated and the Araf side chains can be esterified to ferulic acid (Correia, 2011).

3.4.3.6 Hemicelluloses

Hemicelluloses refer to several heteropolymers present along with cellulose in almost all terrestrial plant cell walls. Hemicelluloses, in general, offer advantages due to their abundance, biodegradability, nontoxicity, and good biocompatibility.[97] Xylans, in particular, are considered for paper coatings and films in food packaging.[98] They are a complex group of hemicelluloses with the common feature of a backbone of β-1,4–linked xylose residues (Figure 3.24).[99]

The main limitation of the industrial application of xylans and hemicelluloses, in general, is their high hydrophilic character, resulting from an abundance of free hydroxyl groups distributed along their backbones and side chains.[100] These hydroxyl groups allow chemical functionalization by attaching hydrophobic groups through, e.g., acetylation.

The variable structure and organization of hemicelluloses requires the combined action of many enzymes, known as hemicellulases, for their complete degradation.[101] The catalytic modules of hemicellulases are either glycoside hydrolases (GHs) that hydrolyze glycosidic bonds or carbohydrate esterases (CEs), which hydrolyze ester linkages of acetate or ferulic acid side groups.

3.4.3.7 Chitosans

Natural polymers must have thermoplastic properties to be successfully processed using existing industrial-scale technologies or even recycling processes. With only a few exceptions, natural polymers are not thermoplastic. However, chemical and physical modification techniques can induce thermoplasticity in them. This is also true for chitosans (Figure 3.25).

Chitosans refer to linear polysaccharides derived from chitin. Chitin is the principal constituent in the exoskeleton of arthropods and the cell walls of fungi; it is the second-largest polysaccharide occurring on the planet. Chitin is a homopolymer of β-(1,4)-linked N-acetyl-D-glucosamine.[102] Chitosan is composed of randomly distributed β-(1,4)–linked D-glucosamine and N-acetyl-D-glucosamine.

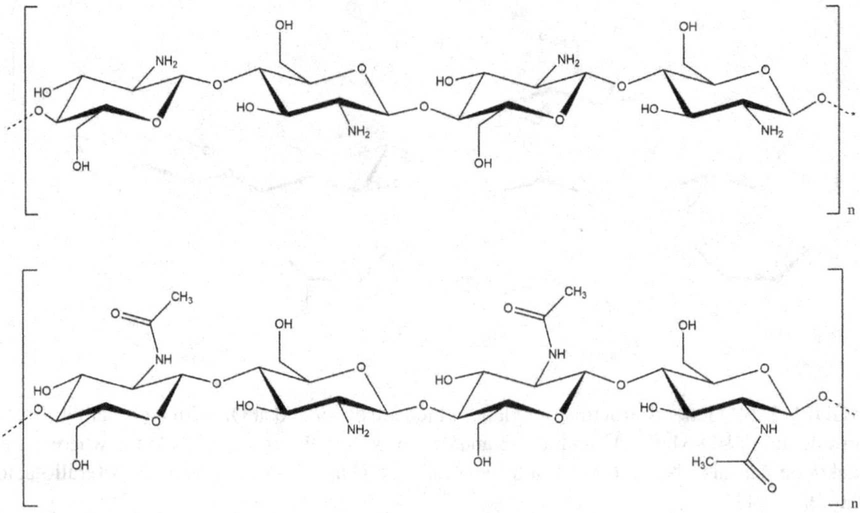

FIGURE 3.25 Structure of completely deacetylated chitosan (top) and randomly deacetylated chitosan (bottom).

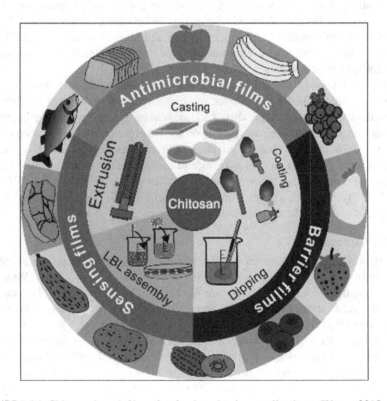

FIGURE 3.26 Chitosan-based films for food packaging applications (Wang, 2018, from *Journal of Agricultural and Food Chemistry* with permission).

Chitosans have a degradation temperature lower than their melting point, which prevents their development in several applications. One way to overcome this issue is the plasticization of the carbohydrates. Biobased plastics include thermoplastic chitosans and modified chitosans obtained using plasticizers such as polyols. Among the nonvolatile polyol plasticizers used to plasticize chitosans, glycerol gives the most important amorphous phase content, while sorbitol gives enhanced mechanical properties.

Chitosan-based materials have been widely applied in various fields for their biological and physical properties of biocompatibility, biodegradability, nontoxicity, antimicrobial ability, and easy film-forming ability.[103] They find applications as antibacterial, barrier, and sensing films (Figure 3.26).

Chitinase and chitosanase enzymes are two members of the glycoside hydrolases family and are characterized by their ability to catalyze the hydrolytic cleavage of chitin and chitosans, respectively.[104]

3.4.3.8 Polysaccharides for Packaging and Especially Edible Films

Polysaccharides are biodegradable, nontoxic, and obtained from renewable resources. They have a high potential as biodegradable packaging material. In packaging applications, polysaccharides used alone or in blends with other polymers include cellulose, starch, chitin, chitosans, pectins, alginates, carrageenans, gum arabic, and gellan gum.[105] Pectins are heterogeneous polysaccharides that are a significant component of the primary cell walls of all land plants, mainly as gel matrices. Alginates, agar, and carrageenans are algal polysaccharides. Gum arabic is a complex plant polysaccharide. Gellan gum is a microbial polysaccharide.

These polysaccharides can be combined with plasticizers, cross-linking agents, antioxidants, and other additives to improve film properties and performance in various food packaging applications. Other films/coating materials include proteins (soy, whey, wheat gluten, gelatin) and lipids. However, the films/coatings prepared with a single material have many deficiencies, which a combination of composite films/coatings can make up. Several prepared methods (electrospinning, casting, extrusion, dipping, fluidized-bed, spraying, panning) used for films/coatings diversify the final applications of edible films (Figure 3.27).[106]

Edible films are thin layers of edible materials that can be used to protect food products from moisture, oxygen, microbial contamination, and other factors that affect their quality and shelf life. Polysaccharides are often used to produce edible films due to their biodegradability, biocompatibility, availability, and film-forming properties:

- Cellulose can be modified into derivatives like methylcellulose or carboxymethylcellulose, forming transparent and biodegradable films. These are used in edible films for their film-forming abilities, with high tensile strength and low water vapor permeability but low solubility and flexibility.[107]
- Chitosans have antimicrobial, antioxidant, and wound-healing properties, and they can form strong and flexible films with low water vapor permeability.

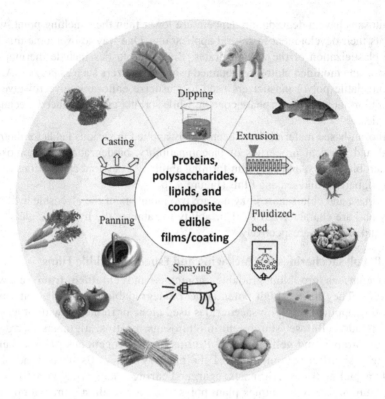

FIGURE 3.27 Proteins, polysaccharides, and lipids-based edible films/coatings for food packaging (Wu, 2023).

- Starch can form transparent and biodegradable films with good mechanical properties but has high water vapor permeability and low water resistance.
- Alginates have gel-forming, emulsifying, and thickening properties, and they can form edible films and coatings with a good oxygen barrier and moderate water vapor barrier. Alginate films, often combined with other polysaccharides, can also incorporate active agents, such as antimicrobials, antioxidants, or enzymes.
- Carrageenan-based formulations are used for edible film preparation. There are three main types of carrageenans with different gel-forming abilities. The gelling ability of pure kappa and iota carrageenans forms excellent films. Pure carrageenan-based coatings are used to form packaging materials for foods.[108]
- Agar can form films with good mechanical strength.
- Pectins have gelling, stabilizing, and thickening properties, and they can form edible films and coatings with a good oxygen barrier and moderate water vapor barrier. Pectin films can also enhance the sensory and nutritional quality of food.
- Gum arabic forms flexible films and is often used as an emulsifier.
- Gellan gum is used in edible films due to its gelling and film-forming properties.

3.4.4 BIODEGRADABLE AND FOSSIL-BASED

Polyesters are the most acceptable group of plastics among biodegradable plastics. All polyesters are theoretically considered biodegradable, as the process used for their manufacturing, i.e., esterification, is a chemically reversible process and can easily be reversed with hydrolytic enzymatic action.[109] Polyesters can be aliphatic, aromatic, or mixed co-polyesters based on the types of monomer. Aliphatic polyesters have linear chain structures, are readily biodegradable, and have low mechanical properties. Aromatic polyesters have enhanced mechanical properties but high resistance to biodegradation. Co-polyesters have properties from both classes, which keep them biodegradable. Polyesters can be biobased, such as PLA and PHA, or fossil-based, like polycaprolactone (PCL) or polybutylene adipate terephthalate (PBAT).

3.4.4.1 Polyesters

PCL is a synthetic, semi-crystalline, biodegradable polyester with a melting point of about 60 °C and a glass transition temperature of about −60 °C (Figure 3.28).

PCL is prepared by ROP of ε-caprolactone using a catalyst and belongs as a polyvinyl alcohol to the group of biodegradable fossil-based polymers.[110] It is entirely degradable through enzyme activities. PCL is highly hydrophobic and thus has longer degradation times than PLA.

PBAT is a biodegradable random copolymer, specifically a copolyester of adipic acid, 1,4-butanedio,l and terephthalic acid.[111]

PBAT is synthesized from a polyester of 1,4-butanediol and adipic acid (BA) and a polyester of dimethyl terephthalate and 1,4-butanediol (BT). Tetrabutoxytitanium is used to catalyze the transesterification of the polyesters of adipic acid and

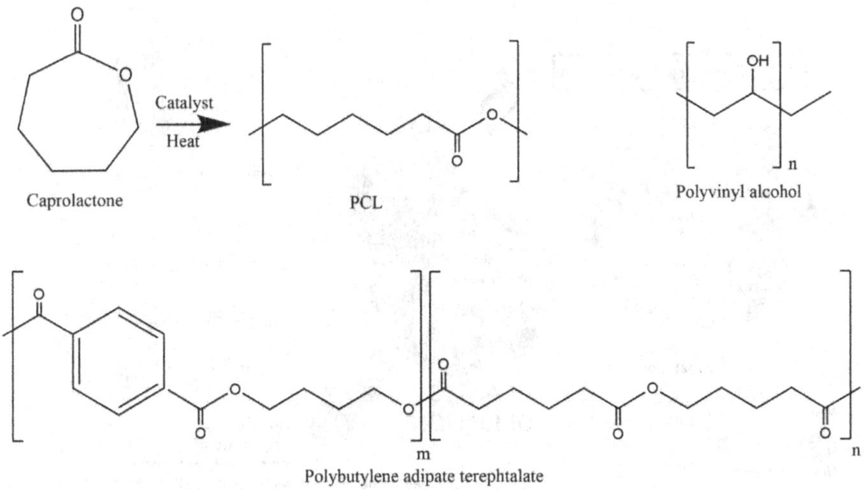

FIGURE 3.28 Polymerization of polycaprolactone (PCL) by ring-opening polymerization of ε-caprolactone using a catalyst.

dimethyl terephthalate to generate the random copolymer PBAT. Particular applications include cling wrap for food packaging, compostable plastic bags for gardening and agricultural use, and water-resistant coatings for other materials, such as paper cups. The biodegradation of PBAT can be regarded as the hydrolysis under the effect of microbial enzymes, during which the BA structure with a noncrystalline portion degrades faster than the BT structure with a crystalline portion.[112]

3.4.4.2 Polyvinyl Alcohol

PVA is a synthetic water-soluble and biodegradable polymer generally prepared by polyvinyl acetate saponification.

PVA is the only fully synthetic C-chain polymer biodegradable in aerobic and anaerobic conditions. However, in laboratory trials, acclimation to allow the growth of selected microorganisms is usually essential to achieve rapid degradation.[113]

PVA is used in a wide variety of industries. It forms strong, flexible films that perform well as oxygen and aroma barriers.[114] It is also resistant to oil, grease, and solvents. These properties make it ideal for sachets, especially if the liquid inside is to be dissolved in water (for example, "pod" packaging for laundry detergent). PVA is frequently used in tablet capsules also.

3.4.5 ALGAE

Plastic biodegradation by algal enzymes and bioplastic production using algal biomass are two new approaches to mitigate plastic pollution (Figure 3.29).[115,116]

On the one hand, algae, particularly microalgae, can degrade plastics through their enzyme system while using plastic polymers as carbon sources for growth. On

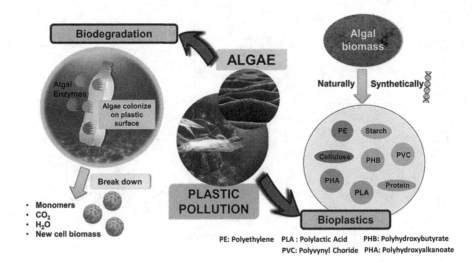

FIGURE 3.29 Plastic biodegradation and bioplastics production using algae (Chia, 2023).

the other hand, bioplastics can be made using protein- and carbohydrate-based polymers from algae. The production of bioplastics is enhanced through genetic engineering and blending.

3.4.6 ARE BIODEGRADABLE PLASTICS A REASONABLE SOLUTION?

Biodegradable plastics are a type of plastic that microorganisms can break down into natural substances, such as water and compost. They are often seen as an eco-friendly alternative to conventional plastics, which can persist in the environment for a long time, cause harm to wildlife and ecosystems, and provide a solution to the environmental issues generated by traditional plastics. However, biodegradable plastics are not a perfect solution to the plastic pollution problem, as they have some drawbacks and limitations.

Ideally, biodegradable plastics (biobased or fossil-based) should be designed to be recyclable or compostable.[117] Unfortunately, current biodegradable plastics pose problems to our waste management system. Indeed, they often end up contaminating the recycling process or the compost bins.[118,119]

The effectiveness and environmental benefits of biodegradable plastics depend on various factors, and their impact can be nuanced. Here are some considerations regarding them:

- Degradation conditions: Biodegradable plastics typically require specific conditions, such as high temperatures, adequate moisture, and specific microbial environments, to break down effectively. If these conditions are not met, they might persist in the environment like conventional plastics.
- Timeframe of degradation: Some biodegradable plastics break down relatively quickly, while others might take longer. In specific environments like landfills, where oxygen and sunlight are limited, degradation might be slow or nonexistent.
- Byproducts of degradation: The breakdown of biodegradable plastics can produce smaller particles or microplastics, which might still pose environmental risks. Additionally, the byproducts released during degradation might not always be environmentally friendly.
- Impact on recycling: Biodegradable plastics can contaminate recycling streams if they are improperly separated. If these plastics enter standard recycling systems, they can reduce the quality of recycled plastic or render it unusable.[120]
- Impact on composting: Compost contamination occurs when noncompostable items are mixed with organics in a compost bin.[121] This trash can be incorporated into finished compost and added to soil. Clear labeling is needed to avoid mistakes, particularly for biodegradable packaging and plastic bags.
- Consumer behavior: There might be confusion among consumers regarding the disposal of biodegradable plastics. Improper disposal can lead to these plastics ending up in the wrong waste streams, reducing their chances of effective degradation.

- Resource requirements: Biodegradable plastics often require specific raw materials and manufacturing processes, which might have their own environmental impact and resource requirements.

Biodegradable plastics should not be considered as the sole solution to the problem of plastic pollution. Careful consideration of their use and disposal is needed to maximize their environmental benefits. They can be helpful in some specific applications, such as food packaging or medical devices, where they can reduce the environmental and health impacts of plastic waste. Ending plastic pollution will happen through reducing overall plastic consumption, promoting reusable alternatives, improving waste management systems, investing in sustainable materials, and adopting a circular economy approach where plastics are designed to be reused, recycled, or composted and where plastic waste is minimized and prevented.

3.5 UPSTREAM PREVENTION SOLUTIONS

Although mismanaged plastic waste is still the primary source of marine plastic pollution, more plastic may be released into the ocean from our human activities like washing and driving than from the mismanagement of our waste. How we currently design, produce, and consume plastic appears highly unsustainable. Preventing plastic pollution at its source remains the best solution to combat it.

3.5.1 Synthetic Textiles

Plastic microfiber pollution produced by domestic and commercial laundering of synthetic textiles has been incriminated as the main source of primary ocean microplastics.[122] PET is the most common microfiber encountered.

Methods for the upstream reduction of microfiber release from laundering mainly fall into two categories:

- Changes to washing machines or washing procedures; modern washing machines need to be fitted at the factory with suitable filters to avoid microfiber release to the waste pipe. Already, innovative solutions are being proposed by large machine-washing manufactories as well as from startup companies.[123]
- Changes in the textiles themselves; avoiding loose knits and fleeces can reduce microfiber release during laundering; new finishing treatments developed by the textile industry are designed to reduce microfiber release.

3.5.2 Tires

When passenger cars, buses, and lorries drive on the road, their tires shed microplastic particles, which pollute the air and are washed into streams, rivers, and the ocean.[124] Tire wear particle (TWP) emissions are gaining more attention since they are considered to contribute the second-largest share to the overall microplastic emissions and are suspected to be harmful to the environment and humans.[125]

In 2021, the EU adopted the Zero Pollution Action Plan, which includes a target to reduce microplastic pollution by 30% by 2030. To meet this objective, strong tire emissions legislation is needed.

Tire industry representatives are working alongside the European Commission at the UN's Task Force on Tire Abrasion to develop standardized testing and tire abrasion limits. In November 2022, the European Commission released a proposal for vehicle emission standards, legislation called Euro 7, which seeks to establish limits for tire particle emissions for the first time globally. Euro 7 offers an opportunity to set a comprehensive framework to tackle tire particle pollution by requiring the industry to phase out the worst-performing tires from the market.[126] In November 2023, the European Parliament supported new rules (Euro 7) to reduce pollutant emissions, including tire abrasion limits. Some innovations, albeit far from solving the global issue, are being developed.[127]

3.5.3 Marine Coatings

The paints applied to commercial ships have been identified as a source of microplastics present as polymers in most anticorrosive and antifouling marine coatings.[128] Furthermore, in-water cleaning operations may amplify the release of microplastics from coatings to remove biofouling.

Graphene and carbon nanotube-based coatings are recognized for their proven antimicrobial and antiadhesive properties. They are environmentally friendly coatings that protect submerged surfaces from the settlement of fouling organisms in marine settings.[129]

3.5.4 Road Markings

Road markings are often listed among meaningful contributors to pollution with microplastics.[130] Resins contained in road markings could be released upon abrasion as microplastics. Effective methods of eliminating the risk of microplastics from this source include promoting road marking materials with extended durability and implementing maintenance programs.[131]

3.5.5 Personal Care Products

In September 2023, the commission adopted measures to restrict intentionally added microplastics.[132] The new rules will prohibit the sale of microplastics as such and in certain consumer products, such as cosmetics and detergents.[133] The restriction covers synthetic polymer particles below 5 millimeters that are organic, are insoluble, and resist degradation. It applies to microplastics present in concentrations greater than 0.01%.

The first measures, for example, the ban on loose glitter and microbeads, will apply on 17 October 2023.[134] In other cases, the sales ban will apply after a more extended period to give affected stakeholders the time to develop and switch to alternatives.

In the United States, the Microbead-Free Waters Act prohibits the manufacturing, packaging, distributing, and selling of rinse-off cosmetic products that contain intentionally added plastic microbeads.[135] According to the FDA, toothpaste is included in the definition. Several U.S. states have also enacted legislation that restricts (e.g., 1 ppm by weight) or bans microbeads in some types of cosmetics and personal care products.

3.5.6 PLASTIC PELLETS

Plastic pellets are one of the largest sources of unintentional microplastic pollution.[136] In October 2023, the European Commission proposed measures to prevent microplastic pollution from unintentionally releasing plastic pellets.[137] Currently, between 52,000 and 184,000 tonnes of pellets are released into the environment each year due to mishandling throughout the entire supply chain. This proposal aims to ensure that all operators handling pellets in the EU take the necessary precautionary measures. This is expected to reduce pellet release by up to 74%, leading to cleaner ecosystems, plastic-free rivers and oceans, and reducing potential risks to human health.

The Federal Pellet Free Waters Act was introduced in the United States in 2022.[138] It is estimated that 230,000 tons of pellets pollute the marine environment yearly. The Plastic Pellet Free Waters Act requires the Environmental Protection Agency to issue a final rule prohibiting certain discharges of plastic pellets and other preproduction plastic into the waters of the United States.

3.5.7 CITY DUST

City dust, which accounts for a large part of ocean microplastics, comes from various sources, including losses from abrasion of objects and infrastructure, weathering of plastic materials, and detergent use.[139] Each is a minor contributor, and a significant reduction requires many individual measures.

3.6 PUBLIC AND CONSUMER AWARENESS

Public awareness can help change how plastic is viewed, used, and managed as waste. Education and engagement can be part of a city's strategic action plan, including consumer awareness campaigns, business awareness campaigns, documentary films, school initiatives, and clean-up activities. The aim is to increase public understanding and shape community perceptions of the dangers of plastic pollution and available solutions, thereby empowering more people and organizations to take action. Community actions can include changes in individual attitudes and purchasing habits, increased sorting and recycling behavior, and responsible business processes and practices. Some various initiatives and campaigns aim to raise public awareness about the use of plastics and encourage widespread action and individual behavioral changes.

3.6.1 Raising Public Awareness

These are examples of how public awareness about the use of plastics is being raised and promoted.

Plastic Smart Cities is a global initiative that helps cities and tourist destinations reduce plastic pollution.[140] It provides tools and resources to help local communities implement plastic waste management strategies, such as public awareness campaigns, business awareness campaigns, documentaries, school initiatives, and clean-ups. One example is the Con Dao Public Awareness Project, which involved a series of workshops, exhibitions, and events to educate residents and visitors to the Con Dao Islands in Vietnam about the dangers of plastic pollution and the benefits of reducing plastic consumption.[141]

Other examples are the Brussels Zero Waste Challenge; #NoPlasticChallenge, France; Parley Ocean Schools; and Plastic Discloser Project,[142] enabling manufacturers, services, and municipalities to manage and reduce their plastic waste by measuring and understanding their plastic footprint.

Earth Day[143] is a global movement that mobilizes people to take action for the environment. One of its campaigns is End Plastic Pollution, which helps people understand the impacts of plastic pollution on human and ecosystem health and how everyday actions can reduce the problem.[144] It provides resources and tools to help individuals, communities, businesses, and governments reduce their plastic footprint, such as the Plastic Pollution Calculator, the Plastic Pollution Primer and Action Toolkit, and the Plastic Pollution Quiz.

Fondation Veolia[145] supports projects addressing environmental and social challenges like plastic pollution. It partners with other foundations, such as Fondation Tara and Fondation Prince Albert II de Monaco, to support research and innovation on ocean plastic pollution and to raise public awareness through expeditions, exhibitions, and calls for projects. One of the examples is the Tara expeditions, which provided the opportunity to measure plastic and microplastic pollution in the Mediterranean and to conduct public awareness campaigns on the subject.[146]

Seas At Risk[147] is the largest umbrella organization of European marine conservation NGOs, promoting ambitious policies at European and international levels. Seas At Risk aims to make seas and oceans abundant in marine life, diverse, climate resilient, and not threatened by human pressures.

3.6.2 Individual Behavioral Changes

Citizens play a crucial role in tackling plastic pollution through individual actions, collective efforts, and advocacy. Some actions taken by citizens include:

- Reducing Single-Use Plastics: Individuals can minimize their use of single-use plastics like plastic bags, straws, bottles, and utensils by choosing reusable alternatives such as cloth bags, metal straws, refillable water bottles, and bamboo utensils.

- Proper Waste Disposal: Ensuring proper disposal of plastic waste by recycling whenever possible and correctly disposing of plastic items in designated recycling bins or centers.
- Participating in Clean-Up Activities: Engaging in beach clean-ups, river clean-ups, and community-led initiatives to remove plastic debris from natural environments.
- Spreading awareness about plastic pollution among friends, family, and communities to highlight its environmental impact and encourage others to adopt more eco-friendly practices.
- Supporting Sustainable Products and Brands: Choosing products with minimal or eco-friendly packaging, supporting brands committed to reducing plastic use, and opting for products made from recycled materials.
- Encouraging Policy Changes: Citizens can advocate for policy changes by supporting petitions, writing to policymakers, and participating in campaigns to implement regulations to reduce plastic pollution.

3.6.3 COMPANY INITIATIVES

3.6.3.1 Coca-Cola

In 2009, Coca-Cola disrupted the sustainable packaging landscape with the launch of PlantBottle, the world's first fully recyclable PET plastic bottle made partially from plants.[148] The PlantBottle includes MEG from sugarcane, but the PTA has been fossil-based until now. In 2023, Coca-Cola unveiled a PET bottle made from 100% plant-based material, excluding the cap and label, using technologies ready for commercial scale.[149]

Coca-Cola's new prototype is made from plant-based paraxylene (bPX)—using a new process by Virent based on sugar from corn—which has been converted to plant-based terephthalic acid (bPTA).

The environmental benefit of non-biodegradable bioplastics is that, because they come from a plant, a certain amount of carbon dioxide is captured during the production of their raw material, positively impacting greenhouse gas emissions.[150] In this regard, a life cycle analysis confirmed that Coca-Cola's Plant-Bottle reduced the CO_2 impact by 12–19% compared to petroleum-based PET plastic bottles.[151] In 2010 alone, the use of this breakthrough packaging eliminated the equivalent of almost 300,000 tonnes of CO_2 or approximately 60,000 barrels of oil. In 2018, Coca-Cola announced its World Without Waste commitment, aiming:

- To collect and recycle the equivalent of one bottle for every bottle sold by 2030 (Collect goal)
- To make 100% of its packaging recyclable by 2025 and to use at least 50% recycled content in its plastic bottles by 2030 (Design goal)
- To unite people through partnerships to support a healthy, debris-free environment (Partner goal).[152]

3.6.3.2 PepsiCo

PepsiCo's 2025 packaging sustainability agenda includes:

- Strive to design 100% of its packaging to be recyclable, compostable, or biodegradable.
- Strive to use 25% recycled content in its plastic packaging by collaborating with its suppliers, helping to increase consumer education, fostering cross-industry and public-private partnerships, and advocating for improved recycling infrastructure and regulatory reform.
- By 2025, PepsiCo will reduce virgin plastic use across their beverage portfolio by 35%, eliminating 2.5 million tonnes of cumulative virgin plastic. The target is based on a 2018 baseline.
- Work to increase recycling rates.[153]

PepsiCo recognizes that its plastic packaging has caused significant concern. In 2018, PepsiCo used 2.3 million tonnes of plastic to package products throughout its food and beverage portfolio. PepsiCo's sustainable plastics vision is to build a world where plastics need never become waste. They aim to achieve that vision through reducing, recycling, and reinventing.

3.6.3.3 McDonald's

In response to the pressing issue of plastic pollution, McDonald's has demonstrated its commitment to environmental stewardship by adopting a zero-plastic strategy.[154] McDonald's commitment to reducing plastic waste through innovative approaches and sustainable practices reflects its dedication to mitigating plastic pollution and preserving the environment.

The five focus areas of its strategy are:

- Eliminating unnecessary packaging and streamlining materials for easier recovery
- Transitioning away from virgin fossil fuel-based plastics
- Increasing the use of recycled materials
- Advancing a circular economy
- Partnering to increase the scale of solutions[155]

3.6.3.4 Shell

Shell is a leading member of the new Alliance to End Plastic Waste, committed to helping end plastic waste in the environment.[156] The alliance of global companies also involves the World Business Council for Sustainable Development.[157]

Shell is concerned about global plastic waste. It supports the need for improved circularity of global plastics markets—encouraging reduction, reuse, and recycling where possible to mitigate plastic release into the environment. To reduce plastic waste, Shell is:

- Investing and innovating in technology that uses plastic waste to create new products and fosters a circular economy

- Collaborating with industry partners, customers, and others to support solutions that tackle and stop plastic waste leakage
- Finding ways to reduce, reuse, and recycle plastic waste in areas of its supply chain

3.6.3.5 TotalEnergies

Since 2019, TotalEnergies has been a founding member of the Alliance to End Plastic Waste, bringing together more than 90 companies, civil society groups, and organizations across the plastics and consumer goods value chain.[158] Members have committed to spending $1.5 billion within five years on solutions to eliminate plastic waste in the environment, particularly in the oceans.

By transforming oil, gas, biomass, and plastic waste, TotalEnergies produces polymers, recycled polymers, and biopolymers that can be incorporated into the plastic manufacturing process. TotalEnergies has a role in improving how end-of-life plastics are managed, thereby preventing them from ending up in the natural environment. The company spearheads several projects to reduce the environmental impact of its polymers, promote recycling, and produce polymers from renewable materials. It is aiming to make 30% recycled and renewable polymers by 2030.

3.6.4 THE OCEAN CLEANUP INITIATIVE

To effectively solve the problem of plastic pollution, we need to turn off the tap and mop the floor simultaneously.[159] The Ocean Cleanup, a nonprofit organization, is developing and scaling technologies to rid the world's oceans of plastic. To achieve this objective, the organization has to work on closing the sources of plastic pollution and cleaning up what has already accumulated in the ocean.

The Ocean Cleanup is cleaning up the garbage patches (Figure 3.30).[160,161]

The fundamental challenge of cleaning the ocean's garbage patches is that the plastic pollution is highly diluted, spanning millions of square kilometers. Their clean-up solution is first designed to concentrate the plastic to collect and remove vast quantities. To clean the oceans, it is also necessary to stop new trash from flowing into them. Rivers are the primary source of ocean plastic pollution. They are the arteries that carry waste from the land to the ocean. According to the organization, 1000 rivers are responsible for roughly 80% of riverine pollution. By tackling this problem, 80% of riverine pollution reaching the ocean can be halted. However, recently the initiative suffered controversy, as experts were doubting the cleaning capacities of the organization, claiming the photos from the supposed event looked staged (Figure 3.30).[162]

3.7 VISION FOR SUSTAINABLE PLASTICS

3.7.1 EUROPE

European plastics manufacturers have agreed on a "Plastics Transition" roadmap to accelerate the transition to make plastics circular, drive life cycle emissions to net zero, and foster the sustainable use of plastics.[163] The roadmap establishes a pathway to reduce greenhouse gas emissions from the overall plastics system by 28% by 2030

FIGURE 3.30 System 002 deployed for testing in the Great Pacific Garbage Patch. The Ocean Cleanup solution consists of two vessels, which slowly pull a long retention system through the plastic-rich areas. Plastic enters the system at the open end, where plastic is then transported along the so-called wing sections into the retention zone, acting like a giant bag (The Ocean Cleanup Project; https://theoceancleanup.com/media-gallery/#&gid=5&pid=4).

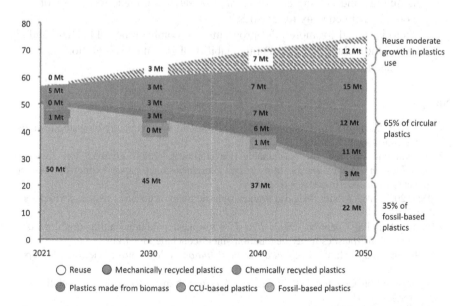

FIGURE 3.31 Circular and fossil-based plastics used by European converters in Mt (Deloitte Analysis, 2023) (Ref Plastics Europe, 2023).

and toward net zero by 2050. In parallel, it predicts the gradual substitution of fossil-based plastics and that circular plastics could meet 25% of European demand in 2030 and 65% by 2050 (Figure 3.31).[164] Circular plastics include mechanically and chemically recycled plastics, plastics from biomass and CCU-based plastics. Cumulated additional investments and operational costs to reach these ambitions are projected to be €235 billion.

European converter's uses are estimated at 57 Mt with 12% of circular plastics in 2021, 60 Mt with 25% in 2030, 62 Mt with 40% in 2040, and 65 Mt with 65% of circular plastics in 2050.[165]

3.7.2 UNITED STATES

To help inform policymakers, business leaders, and the public on ways to advance more sustainable and circular solutions for durable goods at the end of life, the American Chemistry Council's (ACC's) Plastics Division has released an industry roadmap.[166] Its overarching themes include:

- Durable products and their components need to be designed to be disassembled, separated, and repaired or replaced, and spent components need to be recycled and remade into new products.
- The importance of advanced (chemical) recycling can increase the amount and scope of plastics often used in durable applications for recycling and that the mechanical recycling stream cannot easily recycle.
- Standards, methods, and certification programs play an essential role in facilitating and helping ensure durable products are made to be part of a more circular economy for plastics.
- There is a need for more pilot programs that combine the value chain and inform the technical and economic viability of end-of-life separation, sortation, and recycling of durable plastics.

NOTES

1 UNDP, 2023, *Step Towards Ending Plastic Pollution*, www.undp.org/blog/step-towards-ending-plastic-pollution-2040
2 International Institute for Sustainable Development, 2023, *Options for Elements to Include in Plastics Treaty*, https://sdg.iisd.org/news/unep-paper-outlines-options-for-elements-to-include-in-plastics-treaty/
3 IISD, 2023, *INC-3*, https://sdg.iisd.org/news/plastic-treaty-negotiators-request-revised-zero-draft-for-inc-4/
4 Plastic Pollution Coalition, 2023, *INC-3*, www.plasticpollutioncoalition.org/blog/2023/11/20/un-global-plastics-treaty-negotiations-inc-3-conclude-in-nairobi
5 High Ambition Coalition, 2023, *End Plastic Pollution by 2040*, https://hactoendplasticpollution.org/
6 High Ambition Coalition, 2023, *Become Member*, https://hactoendplasticpollution.org/become-member/
7 Oceanic Society, 2023, *7 Solutions to Ocean Plastic Pollution*, www.oceanicsociety.org/resources/7-ways-to-reduce-ocean-plastic-pollution-today/

8 TNO, 2023, *From Plastic Free to Future-Proof Plastics*, www.tno.nl/en/newsroom/2023/06/from-plastic-free-future-proof-plastics/
9 UNEP, 2023, *Turning Off the Tap*, www.unep.org/resources/turning-off-tap-end-plastic-pollution-create-circular-economy
10 PlasticSoup Foundation, 2023, *Mission and Vision*, www.plasticsoupfoundation.org/en/about-us/mission-and-vision/
11 Life Cycle Initiative, 2023, *Life Cycle Approach to Plastic Pollution*, www.lifecycleinitiative.org/activities/life-cycle-assessment-in-high-impact-sectors/life-cycle-approach-to-plastic-pollution/
12 International Institute for Sustainable Development, Earth Negotiations Bulletin, 2013, *INC-2*, https://enb.iisd.org/plastic-pollution-marine-environment-negotiating-committee-inc2-summary
13 Evreka, 2023, *Why is Recycling Sustainable*, https://evreka.co/blog/why-is-recycling-sustainable/
14 AMBR, 2022, *Recycling*, https://ambr-recyclers.org/2022/09/are-my-plastics-really-being-recycled-when-the-answer-is-yes-sometimes-and-never/
15 AMBR, 2022, *Platform*, https://ambr-recyclers.org/2022/09/ambrs-platform-for-improving-recycling-reducing-plastic-consumption/
16 Chemical Recycling Europe, 2023, *About*, www.chemicalrecyclingeurope.eu/
17 Nova-Institute, 2023, *Advanced Recycling Technologies*, https://renewable-carbon.eu/news/the-advanced-recycling-conference-2023-your-ultimate-guide-to-cutting-edge-recycling-solutions/
18 Advanced Recycling Conference, 2023, *Programme*, https://advanced-recycling.eu/program/
19 MILTON ROY, *Overview of Advanced Recycling*, www.miltonroy.com/en-co/resources/blog/overview-of-advanced-recycling
20 Z.O.G. Schyns, 2020, *Macromolecular Rapid Communications*, https://onlinelibrary.wiley.com/doi/full/10.1002/marc.202000415
21 Cefic, *Chemical Recycling*, https://cefic.org/5-things-that-need-to-happen-now-for-chemical-recycling-to-contribute-to-eu-circular-economy/
22 A. Toktarova, 2022, *Journal of Cleaner Production*, Vol. 394, p. 133891, www.sciencedirect.com/science/article/pii/S0959652622034643
23 D. Saebea, 2020, *Energy Reports*, Vol. 6, p. 202, www.sciencedirect.com/science/article/pii/S2352484719306705
24 Cleantech, 2022, *Enzymatic Recycling*, www.cleantech.com/enzymatic-depolymerization-and-recycling-using-enzymes-to-convert-linear-waste-streams-into-circular-supply-chains/
25 NREL, 2022, *PET*, www.nrel.gov/news/features/2022/scientists-discover-enzymes-cheaper-to-recycle-waste-polyester-textiles-and-bottles-than-making-from-petroleum.html
26 Entreprendre, 2023, *PET Recycling*, www.entreprendre.fr/carbios-va-devenir-leader-mondial-du-recyclage-pet-polyethylene-terephtalate-dici-2035/
27 Carbios, *Enzymatic Recycling*, www.carbios.com/en/enzymatic-recycling/
28 Initiatives for the Future of Great Rivers, 2020, *Carbios*, www.initiativesfleuves.org/actualites/carbios-detruire-recycler-plastique-grace-a-enzyme/
29 Carbios, 2022, *Plant in France*, www.carbios.com/en/carbios-to-build-in-france-its-plant/
30 Carbios, 2023, *Novozymes*, www.carbios.com/fr/carbios-et-novozymes-partenariat-strategique/
31 Carbios, 2023, *Plant*, https://renewable-carbon.eu/news/carbios-obtains-building-and-operating-permits-in-line-with-announced-schedule-for-worlds-first-pet-biorecycling-plant-in-longlaville/
32 TIPA, Compostable Packaging, 2023, *Organic Recycling?* https://tipa-corp.com/blog/whats-organic-recycling/
33 Sulapac, 2023, *Compostability*, www.sulapac.com/compostability/

34 M. Xanthos, 1998, "Solvolysis", in *Frontiers in the Science and Technology of Polymer Recycling*, p. 425, https://link.springer.com/chapter/10.1007/978-94-017-1626-0_20

35 Forbes, 2023, *CCU*, www.forbes.com/sites/mariannelehnis/2023/01/11/is-recycling-co2-the-answer-to-decarbonising-the-industrial-sector/?sh=20652e7a75e0

36 Nova Institute, 2023, *CCU*, https://renewable-carbon.eu/news/news-from-ccu/

37 Idtechex, 2023, *CCU*, www.idtechex.com/en/research-article/can-the-plastics-industry-become-a-carbon-capture-leader/27076

38 UNEP, Basel Convention, 2011, *Plastic Waste Amendments*, www.basel.int/Implementation/Plasticwaste/PlasticWasteAmendments/FAQs/tabid/8427/Default.aspx

39 Be the Story Ellen Macarthur, www.be-the-story.com/en/plastic/the-new-plastics-economy-and-the-vision-of-ellen-macarthur/

40 Ellen Macarthur Foundation, 2022, *Press Release*, www.ellenmacarthurfoundation.org/press-release-progress-needs-fresh-acceleration

41 UNEP, 2018, *Exploring Alternative Materials to Reduce Plastic Pollution*, www.unep.org/news-and-stories/press-release/exploring-alternative-materials-reduce-plastic-pollution

42 Science Direct, P. Goswami, 2016, *Biodegradability*, www.sciencedirect.com/topics/chemistry/biodegradability

43 https://en.wikipedia.org/wiki/Biodegradable_plastic

44 Good Start Packaging, www.goodstartpackaging.com/biodegradable-vs-compostable-what-is-the-difference/

45 Y. Tokiwa, 2009, *International Journal of Molecular Sciences*, Vol. 10, p. 3722, www.ncbi.nlm.nih.gov/pmc/articles/PMC2769161/

46 European Bioplastics, *Oxo-Degradable Plastics*, www.european-bioplastics.org/bioplastics/standards/oxo-degradables/

47 Plastic Solutions Review, 2022, *Oxo-Degradable Plastics*, https://plasticsolutionsreview.com/oxo-degradable-plastics/

48 Swiftpak, *Bioplastics*, www.swiftpak.co.uk/insights/biodegradable-vs-compostable-vs-oxo-degradable-the-sustainability-of-bioplastics-explained

49 Normec OWS, *Industrial Composting*, https://normecows.com/degradation-toxicity/industrial-composting/

50 ISO, 2012, ISO 14855-1:2012, *Biodegradability*, www.iso.org/standard/57902.html

51 European Bioplastics, *Compostable Product*, www.european-bioplastics.org/faq-items/what-are-the-required-circumstances-for-a-compostable-product-to-compost/

52 ISO, 2012, *Biodegradability*, www.iso.org/standard/57902.html

53 European Bioplastics, 2022, *Bioplastic Materials*, www.european-bioplastics.org/bioplastics/materials/

54 European Bioplastics, 2023, www.european-bioplastics.org/market/ and Statista, 2023, *Bioplastics*, www.statista.com/statistics/678684/global-production-capacity-of-bioplastics-by-type/

55 M.H. Rahman, 2021, *Journal of Cleaner Production*, Vol. 294, p. 126218, www.sciencedirect.com/science/article/abs/pii/S0959652621004388

56 European Bioplastics, *Recycling*, www.european-bioplastics.org/bioplastics/waste-management/recycling/

57 Institute of Food Technologists (IFT), 2014, *Biobased But not Biodegradable*, www.ift.org/news-and-publications/food-technology-magazine/issues/2014/june/features/biobasedpackaging

58 K. Phung, 2021, *JECE*, Vol. 9, p. 105673, www.researchgate.net/publication/351580016_BioPropylene_Production_Processes_a_critical_review

59 Biokunststofftool, 2023, *Bio-PET*, https://biokunststofftool.de/materials/bio-pet/?lang=en

60 Omnexus, 2023, *PEF*, https://omnexus.specialchem.com/selection-guide/polyethylene-furanoate-pef-bioplastic

61 J.G. Rosenboom, 2022, *Nature*, Vol. 7, p. 117, www.nature.com/articles/s41578-021-00407-8

62 P. Stegmann, 2023, *Journal of Cleaner Production*, Vol. 395, p. 136426, www.sciencedirect.com/science/article/pii/S095965262300584X#bib11

63 Avantium, February 21, 2023, *Press Release*, www.avantium.com/2023/avantium-and-origin-materials-to-accelerate-the-mass-production-of-fdca-and-pef-for-advanced-chemicals-and-plastics/

64 K. Sudesh and T. Iwata, 2008, *Clean*, Vol. 36, p. 433, https://onlinelibrary.wiley.com/doi/10.1002/clen.200700183

65 M. Rafiee, 2021, *AIMS Materials Science*, Vol. 8, p. 524, www.aimspress.com/article/doi/10.3934/matersci.2021032?viewType=HTML

66 TotalEnergies Corbion, 2023, *Planting the Future with PLA*.

67 B.L.C. Cunha, 2022, *Bioengineering (Basel)*, Vol. 9, p. 164, www.ncbi.nlm.nih.gov/pmc/articles/PMC9032396/

68 A.Z. Naser, 2021, *RSC Advances*, Vol. 11, p. 17151.

69 https://polylactide.com/plla-vs-pdla-vs-pdlla/

70 T.M. Ovitt, 2002, *Journal of the American Chemical Society*, Vol. 124, p. 1316, https://pubs.acs.org/doi/10.1021/ja012052%2B#

71 R.M. Michell, 2023, *Progress in Polymer Science*, Vol. 146, p. 101742, www.sciencedirect.com/science/article/pii/S0079670023000965

72 V.H. Sangeetha, 2016, *Polymer Engineering and Science*, Vol. 56, No. 6, p. 669.

73 https://en.wikipedia.org/wiki/Tacticity

74 E. Marshall, *Biorenewable Polymers 1: The Isotactic Polymerisation of Lactide*, Imperial College London.

75 Gianeco, PLA, www.gianeco.com/en/faq-detail/1/10/how-long-does-pla-take-to-decompose

76 E.R. Ghomi, 2021, *Polymers*, Vol. 13, p. 1854, www.mdpi.com/2073-4360/13/11/1854

77 TWI, *PLA*, www.twi-global.com/technical-knowledge/faqs/what-is-pla

78 K.J. Jem, 2020, *Advanced Industrial and Engineering Polymer Research*, Vol. 3, p. 60, www.sciencedirect.com/science/article/pii/S2542504820300026

79 W. Zhou, 2023, *Journal of Environmental Management*, Vol. 341, p. 118033, www.sciencedirect.com/science/article/pii/S0301479723008216

80 K.W. Meereboer, 2020, *Green Chemistry*, https://pubs.rsc.org/en/content/articlelanding/2020/gc/d0gc01647k

81 Sourcegreen, www.sourcegreen.co/plastics/pha-biobased-polymer-packaging/

82 S.Y. Hwang, 2012, *Polymer Journal*, Vol. 44, p. 1179, www.nature.com/articles/pj2012157

83 M. Barletta, 2022, *Progress in Polymer Science*, Vol. 132, p. 101159, www.sciencedirect.com/science/article/abs/pii/S0079670022000776

84 L. Aliotta, 2022, *Polymers*, Vol. 14, p. 844, www.ncbi.nlm.nih.gov/pmc/articles/PMC8963078/

85 S.A. Rafiqah, 2021, *Polymers*, Vol. 13, p. 1436, www.ncbi.nlm.nih.gov/pmc/articles/PMC8125033/

86 M.J. Mochane, 2021, *Polymers*, Vol. 13, p. 1200, www.ncbi.nlm.nih.gov/pmc/articles/PMC8068185/

87 S. Pérez, 2010, *Starch*, Vol. 62, p. 389, The molecular structures of starch components and their contribution to the architecture of starch granules: A comprehensive review, https://onlinelibrary.wiley.com/doi/abs/10.1002/star.201000013

88 Patent, 2010, https://patents.google.com/patent/EP2467418B1/en

89 Z.N. Diyana, 2021, *Polymers*, Vol. 13, p. 1396, www.ncbi.nlm.nih.gov/pmc/articles/PMC8123420/

90 J.L. Wertz, 2023, *Biomass in the Bioeconomy*, CRC Press, www.taylorfrancis.com/books/ mono/10.1201/9781003308454/biomass-bioeconomy-jean-luc-wertz-philippe-mengal-serge-perez

91 R. Hirase, 2022, *Carbohydrate Polymers*, Vol. 275, p. 118669, www.sciencedirect.com/ science/article/abs/pii/S0144861721010560

92 T. Heinze, 2005, *Cellulose Solvents*, www.scielo.br/j/po/a/f3MgHyD9THWMDvzcDWv QSmv/?lang=en#

93 S. Acharya, 2021, *Polymers*, Vol. 13, p. 4344, www.ncbi.nlm.nih.gov/pmc/articles/ PMC8704128/

94 J.L. Wertz, 2010, *Cellulose Science and Technology*, EPFL Press, www.taylorfrancis.com/ books/mono/10.1201/b16496/cellulose-science-technology-jean-luc-wertz-jean-mercier-olivier-b%C3%A9du%C3%A9

95 Elevate Packaging, 2020, *Compostable Cellophane*, https://elevatepackaging.com/blog/ compostable-cellophane/

96 Springer, 2023, *Cellulose Acetate*, https://link.springer.com/article/10.1007/s10924-010-0258-0

97 P. Nechita, 2023, *Polymers*, Vol. 15, p. 2088, www.mdpi.com/2073-4360/15/9/2088

98 P.Nechita,2021,*Sustainability*,Vol.13,p. 13504,www.mdpi.com/2071-1050/13/24/13504

99 M.A.S.Correia,2011,*JBC*,Vol.286,p. 22510,www.jbc.org/article/S0021-9258(19)48960-6/fulltext

100 L. Hu, 2020, *American Journal of Plant Sciences*, Vol. 11, No. 12, www.scirp.org/journal/ paperinformation?paperid=106012

101 A. Sapkota, 2022, *Microbe Notes*, https://microbenotes.com/microbial-degradation-of-hemicellulose/

102 S. Perez, 2021, *Chitin and Chitosans in the Bioeconomy*, CRC Press, www.taylorfrancis.com/ books/mono/10.1201/9781003226529/chitin-chitosans-bioeconomy-jean-luc-wertz-serge-perez

103 H. Wang, 2018, *Journal of Agricultural and Food Chemistry*, Vol. 66, p. 395, https://pubs. acs.org/doi/10.1021/acs.jafc.7b04528

104 X. Zhu, 2007, *Journal of Zhejiang University Science B*, Vol. 8, p. 831, www.ncbi.nlm. nih.gov/pmc/articles/PMC2064955/

105 K.V. Aleksanyan, 2023, *Polymers*, Vol. 15, p. 451, www.ncbi.nlm.nih.gov/pmc/articles/ PMC9865279/

106 Y. Wu, 2023, *Food Biophysics*, https://doi.org/10.1007/s11483-023-09794-7

107 A. Kocira, 2021, *Agronomy*, www.mdpi.com/2073-4395/11/5/813

108 C. Cheng, 2022, *Frontiers in Nutrition*, www.frontiersin.org/articles/10.3389/fnut.2022. 1004588/full

109 S.M. Satti, 2020, *Applied Microbiology International*, https://ami-journals.onlinelibrary. wiley.com/doi/full/10.1111/lam.13287

110 ScienceDirect, *Polycaprolane*, www.sciencedirect.com/topics/medicine-and-dentistry/ polycaprolactone

111 https://en.wikipedia.org/wiki/Polybutylene_adipate_terephthalate

112 Y. Fu, 2020, *Molecules*, Vol. 25, p. 3946, www.ncbi.nlm.nih.gov/pmc/articles/PMC 7504808/

113 O. Alonso-Lopez, 2021, *Polymers*, Vol. 13, p. 3742, www.ncbi.nlm.nih.gov/pmc/articles/ PMC8588384/

114 Source Green, 2022, *PVA*, www.sourcegreen.co/materials/polyvinyl-alcohol-pva-pvoh/

115 W.Y. Cheah, 2023, *Algal Research*, Vol. 71, p. 103078, www.sciencedirect.com/science/ article/abs/pii/S221192642300111X

116 W.Y. Chia, 2020, *Environmental Science and Ecotechnology*, Vol. 4, www.sciencedirect. com/science/article/pii/S2666498420300570

117 C. Gioia, 2021, *Chemistry Europe*, https://chemistry-europe.onlinelibrary.wiley.com/doi/full/10.1002/cssc.202101226

118 Biopak, www.biopak.com/au/resources/biodegradable-plastic-problems

119 Science the Wire, 2023, https://science.thewire.in/environment/bioplastic-recycling-composting-pathways/

120 Environmental Protection Agency, 2023, www.epa.gov/trash-free-waters/frequently-asked-questions-about-plastic-recycling-and-composting

121 Elevate Packaging, 2023, https://elevatepackaging.com/blog/compost-contamination/

122 C. Gaylarde, 2021, *Heliyon*, Vol. 7, p. e07105, www.sciencedirect.com/science/article/pii/S2405844021012081

123 https://planetcare.org/fr; www.gulp.online/; www.youtube.com/watch?v=tm-6gKZVf3s

124 PEW, 2023, *Tyres*, www.pewtrusts.org/en/research-and-analysis/articles/2023/11/06/to-fight-microplastic-pollution-eu-needs-strong-tyre-emissions-legislation

125 D. Mennekes, 2022, *Science of the Total Environment*, Vol. 830, p. 154655, www.sciencedirect.com/science/article/pii/S004896972201748X

126 European Parliament, 2023, *Road Transport Emissions*, www.europarl.europa.eu/news/en/press-room/20231009IPR06746/euro-7-meps-back-new-rules-to-reduce-road-transport-emissions

127 https://thetyrecollective.com/; Michelin - Towards a 100% sustainable tire

128 M.N. Tamburri, 2022, *Frontiers in Marine Science*, p. 1074654, www.frontiersin.org/articles/10.3389/fmars.2022.1074654/full

129 F. Sousa-Cardoso, 2022, *Antibiotics*, Vol. 11, p. 1102, https://pubmed.ncbi.nlm.nih.gov/36009971/

130 T.E. Burghardt, *Transportation Research Part D*, Vol. 102, p. 103123, www.sciencedirect.com/science/article/pii/S1361920921004181

131 European Union Road Federation, 2022, *Road Markings and Microplastics*, https://erf.be/publications/microplastics/

132 European Commission, 2023, *Intentionally Added Microplastics*, https://ec.europa.eu/commission/presscorner/detail/%20en/ip_23_4581
PlanetCare | The Most Effective Solution to Stop Microfiber Pollution
Gulp: Stop microplastic pollution from your laundry—one gulp at a time
How To Install a Microplastic Filter on Washing Machine | Samsung UK (youtube.com)

133 Productwise, 2023, *Intentionally Added Microplastics*, https://products.cooley.com/2023/10/03/eu-adopts-restriction-of-intentionally-added-microplastics/

134 European Chemicals Agency, 2023, *Microplastics*, https://echa.europa.eu/hot-topics/microplastics

135 Compliancegate, 2023, *Microplastics*, www.compliancegate.com/microplastic-regulations-united-states/

136 EUbusiness, 2023, *Microplastics*, www.eubusiness.com/news-eu/microplastics.plastic-pellets.16u/

137 European Commission, 2023, *Pellets*, https://ec.europa.eu/commission/presscorner/detail/en/ip_23_4984

138 Congress, 2022, *Pellets*, www.congress.gov/bill/117th-congress/house-bill/7861

139 First Sentier MUFG, 2023, *Microplastics*, www.firstsentier-mufg-sustainability.com/insight/sources-of-microplastics-and-their-distribution-in-the-environment.html

140 Plastic Smart Cities, 2023, *Public Awareness*, https://plasticsmartcities.org/public-awareness/

141 Vietnam Insider, 2022, *Con Dao Island*, https://vietnaminsider.vn/wwf-commits-to-reduce-plastic-waste-in-con-dao-island/

142 Plastic Disclosure Project, www.plasticdisclosure.org/

143 Earth Day, 2024, *The Official Site*, www.earthday.org/

144 Earth Day, 2024, *End Plastic Pollution*, www.earthday.org/campaign/end-plastic-pollution/
145 Foundation Veolia, www.fondation.veolia.com/en/raising-awareness-about-plastic-pollution-and-encouraging-widespread-action
146 Fondation Veolia, 2019, *Mission Tara Microplastiques*, www.fondation.veolia.com/fr/mission-microplastiques-etudier-les-origines-de-la-pollution-en-remontant-10-fleuves-europeens
147 Seas at Risk, https://map.seas-at-risk.org/wp-content/uploads/2021/06/ENGLISH-Catalogue.pdf
148 Coca-Cola, 2019, *Plantbottle*, www.coca-colacompany.com/media-center/coca-cola-expands-access-to-plantbottle-ip
149 Coca-Cola, 2023, *Plant-Based Bottle*, www.coca-colacompany.com/media-center/100-percent-plant-based-plastic-bottle
150 NCBI NIH, 2023, *Bioplastics*, www.ncbi.nlm.nih.gov/pmc/articles/PMC10032476/
151 Imperial College London, 2019, *Carbon Footprint*, www.imperial-consultants.co.uk/casestudies/reducing-carbon-footprint-of-plastic-bottle/
152 Coca-Cola, 2023, *World Without Waste—Sustainable Packaging*, www.coca-colacompany.com/sustainability/packaging-sustainability
153 PepsiCo, 2024, *Recycling and Sustainability Initiatives*, https://contact.pepsico.com/pepsico/article/pepsico-recycling-and-sustainability-initiatives
154 McDonald's, 2023, *Zero Plastic Strategy*, www.linkedin.com/pulse/from-waste-sustainability-mcdonalds-zero-plastic
155 McDonald's, 2022, *Packaging, Toys and Waste*, https://corporate.mcdonalds.com/corp-mcd/our-purpose-and-impact/our-planet/packaging-toys-and-waste.html
156 Shell, *Plastic Waste*, www.shell.com/sustainability/environment/plastic-waste.html
157 World Business Council for Sustainable Development, www.wbcsd.org/Overview/About-us
158 TotalEnergies, 2024, *Polymers, Polymers and Plastics: The Amount of Recycled and Renewable Materials Keeps Growing*, https://totalenergies.com/group/energy-expertise/transformation-development/polymers
159 The Ocean Cleanup, 2023, https://theoceancleanup.com/
160 Cleaner Oceans Operation, *Boyan Slat's Cleanup Project*, www.oceansplasticcleanup.com/Cleaning_Up_Operations/Boyan_Slat_Ocean_Cleanup_Project_The.htm
161 The Ocean Cleanup Project, https://twitter.com/TheOceanCleanup/status/1450917091469451265/photo/1
162 The Ocean Cleanup Controversy, *Explained*, greenmatters.com
163 Euractiv, 2023, *Plastics Manufacturers Unite Around a Common Vision*, http://pr.euractiv.com/pr/plastics-manufacturers-unite-around-common-vision-redesign-european-plastics-system-256322
164 Deloitte, 2023, *The Plastics Transition*, https://www2.deloitte.com/be/en/pages/climate-and-sustainability/articles/the-plastics-transition.html
165 Plastics Europe, 2023, *The Plastics Transition Executive Summary*, https://plasticseurope.org/de/knowledge-hub/plastics-transitions-roadmap-summary/
166 www.americanchemistry.com/chemistry-in-america/news-trends/press-release/2023/plastics-industry-roadmap-highlights-immense-opportunities-to-remake-durable-goods-at-end-of-life

4 Conclusions

4.1 MAIN FINDINGS OF THE BOOK

4.1.1 Overall Target: Eliminate Plastic Pollution by 2040

A global shift is underway, gaining economic and political momentum toward a shared international goal of eradicating plastic pollution by 2040, specifically targeting the elimination of mismanaged plastic waste. This ambitious objective finds support in a range of private and public initiatives, as detailed in the following sections.

4.1.1.1 Ellen MacArthur Foundation

In 2010, Dame Ellen MacArthur, the fastest solo sailor to circumnavigate the globe, established the Ellen MacArthur Foundation to accelerate the transition to a circular economy. Addressing the prevalent linear model where plastic packaging products are designed for single use, the foundation advocates eliminating all problematic and unnecessary plastic items; innovating to ensure that all plastics we do need are reusable, recyclable, or compostable; and circulating all plastic items we use to keep them in the economy and out of the environment. This approach combats plastic waste and pollution and reduces greenhouse gas emissions (Figure 4.1).

The foundation also advocates that we must change how we design, use, and reuse plastics. The implementation of reuse models plays a crucial part in this, while compostability is a solution for specific applications, contingent on effective collection and composting infrastructures. Alongside voluntary efforts by businesses, governments are essential in establishing robust infrastructures supported by enabling regulatory frameworks. Simultaneously, businesses are responsible for providing sustainable and circular plastics that are either reusable, recyclable, or compostable (Figure 4.2).

4.1.1.2 UN Environment Assembly and UN Environment Programme

In March 2022, the fifth United Nations Environment Assembly (UNEA-5) unanimously endorsed a historic resolution (resolution 5/14) in Nairobi, setting the mandate to create an international legally binding instrument by 2024 to prevent and reduce global plastic pollution comprehensively. This instrument is designed to address the entire life cycle of plastic. Resolution 5/14 assigns the UN Environment Programme (UNEP) the task of convening an Intergovernmental Negotiating Committee (INC) to develop the instrument, with the INC aiming to conclude negotiations by the end of 2024.

Following the adoption of resolution 5/14, the High Ambition Coalition to End Plastic Pollution, comprising several nations, advocated for the establishment of

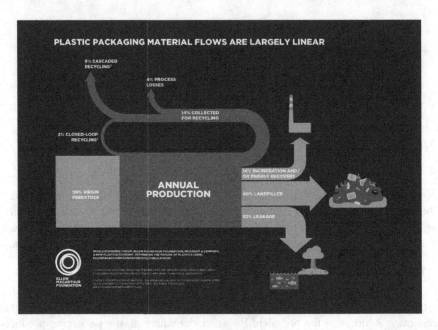

FIGURE 4.1 Largely linear economy for plastic packaging. (Image credits: The Ellen MacArthur Foundation. Please note this graphic cites past data from the 2016 report "The New Plastics Economy: Rethinking the Future of Plastics," with permission.)

FIGURE 4.2 A circular economy for plastic. (Image credit: The Ellen MacArthur Foundation, with permission.)

a robust treaty to end plastic pollution by 2040, safeguard human health and the environment, and contribute to biodiversity restoration while mitigating climate change.[1]

4.1.1.3 Canada and G7 Ministers

In November 2018, federal, provincial, and territorial governments adopted the Canada-wide Strategy on Zero Plastic Waste through the Canadian Council of Ministers of the Environment.[2] This strategy involves the development of targets, standards, and regulations to eliminate plastic pollution in Canada, with the overarching goal of achieving zero plastic waste by 2030.[3,4]

In April 2023, the G7 Ministers of Climate, Energy, and the Environment committed to ending plastic pollution, aiming to reduce additional plastic pollution to zero by 2040. This commitment includes efforts to increase plastics circularity within the economy.[5,6]

4.1.1.4 OECD

In November 2023, the OECD published a report titled "Towards Eliminating Plastic Pollution by 2040: A Policy Scenario Analysis."[7] The report's Global Ambition scenario envisions a near-total elimination of mismanaged waste, with recycling significantly covering 42% of waste generated in 2040. This represents a quadruple increase from the average global recycling rate of 9.5% in 2020.

4.1.1.5 Alliance to End Plastic Waste

Launched in 2019, the industry-founded Alliance to End Plastic Waste (AEPW) strives to develop, deploy, and scale sustainable solutions for minimizing and managing plastic waste, particularly in oceans. By fostering collaboration, innovation, and investment, the AEPW aims to establish a circular economy for plastics, ensuring responsible use, reuse, and recycling.[8] Founding members include BASF, Chevron Phillips Chemical, ExxonMobil, Dow Chemical, Mitsubishi Chemical Holdings, Procter & Gamble, Shell, and TotalEnergies. In November 2022, the Alliance financially supported Indonesia's national goals to reduce ocean plastic pollution by 70% by 2025 and achieve near-zero leakage by 2040. The transition to a circular economy for plastic waste by 2040 is estimated to cost USD 1.2 trillion, according to a Global Plastic Action Partnership report under the World Economic Forum.[9,10]

4.1.1.6 Nordic Council of Ministers

The "Towards Ending Plastic Pollution by 2040" report commissioned by the Nordic Council of Ministers proposes 15 global policy interventions under five pillars: Reduce, Eliminate, Expand Circularity (via reuse, repair, and recycling), Controlled Disposal, and Microplastics (Figure 4.3).[11,12] If supported by common rules in the international legally binding instrument, these interventions could reduce annual volumes of mismanaged plastic waste by 90% and virgin plastic use by 30% by 2040 relative to 2019 levels. The report emphasizes the need for additional measures to

FIGURE 4.3 The 15 global policy interventions under five pillars from the Nordic Council of Ministers (Nordic Council of Ministers, 2023; with permission from NORDEN and Systemiq; Towards Ending Plastic Pollution: 15 Global Policy Interventions for Systems Change. 2023. p. 13. https://doi.org/10.6027/temanord2023-539).

align the plastic system with the Paris Climate Agreement and address health and biodiversity concerns.

4.1.1.7 Oxford University

In an article published in *Nature* in January 2024, scientists from Oxford University[13] outlined a scenario for a circular carbon and plastics economy while achieving net-zero greenhouse gas emissions. This particularly bold scenario includes four targets:

- Fifty percent reduction in plastics demand
- Elimination of fossil-based plastics
- Ninety-five percent recycling rate
- Use of renewable energy

The authors emphasize that the problem is solvable but requires concerted actions across all four target areas in the frame of a UN international plastics treaty.

4.1.2 HOW TO REACH THE TARGET

In a circular economy, the concept of waste is redefined. To effectively eliminate plastic waste by 2040 within a circular economy framework, it's imperative to adhere to the "3R" principle: reduce, redesign, and recycle. In this context, "refuse" is encompassed within "reduce," while "reuse" is integrated into "redesign." These practices align directly with the United Nations' Sustainable Development Goal 12 (Sustainable consumption and production) and involve leveraging new technologies to reduce the prevalence of unsustainable products.[14]

4.1.2.1 Reduce

Reliance solely on recycling will not suffice to address the plastic crisis. The INC is currently debating targets for reducing plastic production, a discussion closely linked to the reuse, recycle, and reduce principles.

Stakeholders such as the plastics industry and oil and petrochemical exporters, including Russia and Saudi Arabia, advocate for a global agreement emphasizing plastic recycling and reuse. However, environmental activists and certain governments argue for a significant reduction in initial production.[15] The European Union has emerged as a leader in global plastic action, which is evident in its support for initiatives like the High Ambition Coalition (HAC) to End Plastic Pollution.[16]

The HAC advocates for a robust treaty addressing the entire plastic life cycle to achieve the goal of ending plastic pollution by 2040. Its statement to INC-3 emphasizes the need for binding provisions within the treaty to curtail and reduce the consumption and production of primary plastic polymers to sustainable levels. This includes eliminating and restricting unnecessary, avoidable, or problematic plastics and their constituents, guided by their adverse environmental and human health effects. The HAC also emphasizes the importance of enhancing the safe circularity of plastics in the economy, managing plastic waste responsibly, and eliminating the release of plastic, including microplastics, into the environment.

Organizations like Greenpeace urge governments to slash plastic production by at least 75% by 2040,[17] emphasizing the critical role of this reduction in mitigating climate change. Meanwhile, the Nordic Council of Ministers suggests that a 30% reduction in virgin plastic use by 2040 relative to 2019 levels would be a significant but achievable target.

The proliferation of single-use plastics is alarming, constituting 36% of total plastic production, with a staggering 85% ending up as waste. Notably, 98% of these single-use plastics are derived from fossil fuels, significantly impacting the global carbon budget.[18,19] Integral components of a circular economy for plastics include the elimination of polymers and chemicals of concern, as well as problematic and avoidable plastic products.[20] This entails phasing out single-use plastic packaging and avoiding single-use plastic products such as bags and straws.[21]

4.1.2.2 Redesign

The overall target will never be reached without redesigning plastics. This action will require research and innovation. Alternative materials and upstream preventive policies can offer solutions toward ending plastic pollution.

4.1.2.2.1 Alternative Materials

Transitioning from disposable plastics to sustainable alternatives is essential for safeguarding the environment. In a circular economy, plastics are designed for reuse, recycling, or composting, thereby minimizing and preventing plastic waste. To facilitate this transition, single-use plastics should be minimized.[22]

The review in Chapter 3 underscores the competitive advantages of biobased polymers over nonrenewable fossil-based polymers, whether biodegradable or not.[23] Biobased polymers save fossil resources by utilizing biomass, are carbon neutral—or at least have a lower carbon footprint—and contribute to a circular economy. Additionally, biodegradability reduces the negative impact of microplastics in the environment. While nonbiodegradable plastics may lead to persistent microplastics, fragments and microplastics from inherently biodegradable materials are more likely to undergo further biodegradation and, ultimately, mineralization.[24]

Key considerations for the growth and design of biobased polymers include their end-of-life pathway:

- Biobased polymers identical to their fossil counterparts, such as bio-PE and bio-PET, can be mechanically recycled in existing recycling streams.
- PLA undergoes slow degradation in nature over a period of 6–24 months.[25] Chemical depolymerization is employed for the advanced recycling of PLA, resulting in lactic acid.
- PHA exhibits versatility in its end-of-life options. It can be reused, recycled back to the polymer for new applications,[26] or biodegraded within 20–45 days under suitable environmental conditions.[27]
- Polysaccharide-based materials, although generally easily biodegradable and compostable, require optimal end-of-life pathways to maximize the circular economy. Starch-based bags, for instance, are commonly composted or subjected to anaerobic digestion, while thermochemical recycling is an alternative for certain applications[28,29]

4.1.2.2.2. Upstream Measures to Combat Microplastics

Upstream measures target the prevention of microplastic proliferation (particles smaller than 5 mm) in the environment, addressing growing concerns regarding their impact on marine ecosystems and human health.

Primary microplastics are directly released into the environment as small particles (15–31% of ocean microplastics).[30] Their main sources are the laundering of synthetic clothes (35% of primary microplastics) and abrasion of tires (28%), as well as intentionally adding microplastics in products such as cosmetics and detergents. Secondary microplastics originate from the degradation of larger plastic objects, such as plastic bags, bottles, or fishing nets (69–81% of microplastics in the oceans.)

The EU has taken a proactive stance in regulating microplastics,[31] adopting restrictions on intentionally added microplastics in products and aiming to establish a harmonized methodology for monitoring microplastic pollution in marine environments. Similarly, the United States passed the Microbead-Free Waters Act, prohibiting the production and sale of personal care products containing microbeads, while Canada listed microbeads as toxic substances. Measures to minimize microplastic release from textiles, tires, marine coatings, road markings, personal care products, plastic pellets, and city dust are also prioritized to combat microplastic pollution.

4.1.2.3 Recycle

The urgent imperative lies in boosting recycling rates and enhancing collection and sorting methods, all while minimizing the use of single-use plastics, which presently constitute half of plastic production.[32] Globally, a mere 9–10% of plastic waste undergoes recycling, with 19% being incinerated, 50% destined for landfills, and 22% improperly managed.

The OECD's Global Ambition scenario envisions a near-complete eradication of mismanaged waste by 2040, with recycling expected to be pivotal. Projections suggest a swift ascent, with recycling encompassing 42% of waste generated by 2040 and potentially reaching 58% with the same policy package but extending to a 2060 target for eradicating mismanaged waste (Figure 4.4).[33]

This surge in recycling rates represents a fourfold increase from current levels. However, limitations in both existing recycling technologies and the availability of suitable materials constrain the expansion of recycling rates. To realize the objectives of this policy scenario, substantial enhancements in recycling processes and reductions in recycling losses are imperative. This necessitates substantial investments in recycling technologies and the adoption of improved designs conducive to recycling.

Economic instruments geared toward bolstering recycling should encompass:

- Landfill and incineration taxes, ensuring that landfilling costs exceed incineration costs and incineration costs exceed recycling costs.

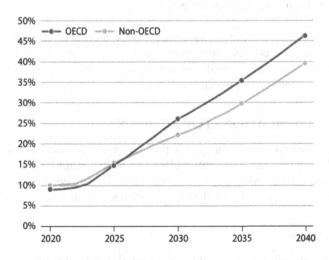

FIGURE 4.4 Recycling rate in percentage of total waste management according to the OECD Global Ambition scenario (OECD, 2023).

- Modulated extended producer responsibility (EPR) schemes for packaging and durables, mandating that companies assume responsibility for product collection, sorting, and recycling in line with the "polluter pays" principle.
- Deposit-refund schemes impose an additional fee upon product purchase, which is refundable upon return of the product's packaging for recycling or reuse.
- Pay-as-you-throw schemes, wherein residents are charged for municipal solid waste collection based on the quantity disposed.

Collectively, these measures serve to incentivize recycling while aligning economic incentives with sustainable waste management practices.

4.1.2.4 Complementary Policies

The Nordic Council of Ministers highlights several complementary policy measures aimed at addressing plastic waste comprehensively:

- **Export Restrictions:** Implementing restrictions on the trade of plastic waste to regions with limited capacity or resources prevents the offloading of waste onto areas ill-equipped to manage it responsibly.
- **Global Standards for Controlled Disposal:** Establishing global standards for controlled disposal ensures that plastic waste that cannot be effectively prevented or safely recycled is managed responsibly and in accordance with established guidelines.
- **Removal Programs for Legacy Plastic:** Initiating removal programs targeting legacy plastic in the environment is crucial for addressing existing pollution. Efforts such as beach and ocean cleanups aid in physically removing plastic waste and play a vital role in raising public awareness and preventing further waste mismanagement.

By implementing these complementary policy measures in conjunction with efforts to reduce, recycle, and redesign plastic usage, governments and stakeholders can work toward achieving significant progress in combating plastic pollution and promoting sustainable waste management practices on a global scale.

4.2 RECOMMENDATIONS FOR ACTIONS

A systemic transformation in our plastic production and consumption approach can only be achieved through collaborative efforts involving citizens, policymakers, academia, and industry, all working toward a common goal.[34] This necessitates coordinated action across three key levels: legislation, research and development, and education.

4.2.1 LEGISLATION

The primary objective is to develop an international legally binding instrument addressing plastic pollution, particularly in marine environments.[35] This instrument must adopt a comprehensive approach encompassing the entire life cycle of plastic, from production and design to disposal. Negotiations for this global agreement are anticipated to culminate in a diplomatic conference in 2025.[36,37]

Furthermore, legislative measures should encompass:

- Banning unnecessary and hazardous plastics
- Promoting the establishment of infrastructures for effective collection, sorting, recycling, and composting
- Restricting the use of unrecyclable materials
- Mandating target percentages of recycled plastic in products
- Implementing taxes or bans on incineration and landfilling of plastics
- Promoting modulated EPR schemes for both single-use and durable plastics
- Advocating for controlled disposal practices
- Supporting cleanup programs to address existing pollution
- Regulating waste trade to regions with limited waste management capacities
- Encouraging the utilization of biobased resources and renewable energy
- Mandating the use of design for longevity and design for recycling principles
- Developing clear regulations and financial incentives to transition from niche polymers to large-scale bioplastic market applications

These legislative actions collectively facilitate the transition toward a circular plastics economy, where virgin polymers are derived from renewable or recycled raw materials. Implementing robust legislation addressing all stages of the plastic life cycle can pave the way for a sustainable and environmentally responsible approach to plastic production and consumption.

4.2.2 RESEARCH AND DEVELOPMENT

Research and development (R&D) initiatives drive innovation and facilitate the transition towards sustainable plastic alternatives.

Key R&D actions may include:

- Developing alternative materials and polymers, focusing on designing them for end-of-life scenarios through recycling or composting
- Evaluating and comparing fossil-based and biobased materials' sustainability and environmental impact[38]
- Increasing the compatibility of biobased plastics with existing and new recycling streams
- Leveraging the biodegradability of certain biobased plastics as a viable end-of-life scenario
- Facilitating the "upcycling" of heterogeneous plastic and bioplastic waste into higher-quality materials
- Revising existing standards for identifying (bio)plastics and guidelines for life cycle assessments
- Enhancing recycling processes in terms of both quantity and quality

- Developing materials and designs to reduce microplastics at their sources, including synthetic textiles during washing, tire abrasion during driving, paints, city dust, plastic pellets, road markings, and personal care products

4.2.3 EDUCATION

Educational efforts are essential in fostering public awareness and promoting sustainable behaviors related to plastic consumption and waste management.

These efforts should:

- Increase citizen and public awareness of plastic pollution through environmental courses and guidance on sustainable consumption behaviors[39]
- Provide clearer guidance on managing plastic product waste, including recycling, home composting, and industrial composting
- Reconnect individuals with nature to cultivate a societal shift toward sustainability and environmental stewardship

By investing in research and development and promoting educational initiatives, we can advance toward a more sustainable future where the harmful impacts of plastic pollution are mitigated through innovation and informed decision-making.

NOTES

1 High Ambition Coalition to End Plastic Pollution, 2023, https://hactoendplasticpollution.org/
2 Canada, 2023, *Zero Plastic Waste*, www.canada.ca/en/environment-climate-change/services/managing-reducing-waste/reduce-plastic-waste/canada-action.html
3 Government of Canada, 2023, *Zero Plastic Waste*, www.canada.ca/en/environment-climate-change/news/2023/04/how-can-canada-cut-pollution-recycle-more-plastic-and-track-plastic-products-nationally-consultations-launched.html
4 Greenpeace, 2024, *Ban on Single-Use Plastics*, www.greenpeace.org/canada/en/act/expand-the-single-use-plastics-ban/
5 G7, 2023, www.g7fsoi.org/g7-climate-energy-and-the-environment-ministerial-2023/
6 Euwid, 2023, *End Plastic Pollution by 2040*, www.euwid-recycling.com/news/policy/g7-ministers-pledge-to-end-plastic-pollution-by-2040-170423/
7 OECD, November 2023, *Eliminating Plastic Pollution by 2040*, www.oecd.org/environment/plastics/Interim-Findings-Towards-Eliminating-Plastic-Pollution-by-2040-Policy-Scenario-Analysis.pdf
8 https://en.wikipedia.org/wiki/Alliance_to_End_Plastic_Waste
9 Alliance to End Plastic Pollution, 2022, *Progress Report 2022*, https://endplasticwaste.org/en/our-stories/progress-report-2022
10 Singapore Economic Development Board, 2024, www.edb.gov.sg/en/business-insights/insights/circular-economic-model-and-lots-of-funding-needed-to-solve-plastic-waste-problem.html
11 Norden, 2023, www.norden.org/en/publication/towards-ending-plastic-pollution
12 Systemiq, 2023, *Towards Ending Plastic Pollution by 2040*, Nordic Council of Ministers, www.systemiq.earth/towards-ending-plastic-pollution-by-2040/
13 F. Vidal, et al., 2024, *Nature*, Vol. 626, p. 45, www.nature.com/articles/s41586-023-06939-z#citeas

14 Strathmore University, 2023, *Circular Economy: Business Models for a Sustainable Future*, https://sbs.strathmore.edu/circular-economy-business-models-for-a-sustainable-future/

15 Reuters, 2023, *UN Plastic Treaty*, www.reuters.com/business/environment/un-plastic-treaty-talks-grapple-with-re-use-recycle-reduce-debate-2023-11-19/

16 High Ambition Coalition, 2023, *Statement for INC3*, https://hactoendplasticpollution.org/hac-ministerial-joint-statement-inc3/

17 Greenpeace, 2023, *To Cut Plastic Production*, www.greenpeace.org/international/story/62928/why-greenpeace-is-calling-on-governments-to-cut-plastic-production-by-at-least-75-by-2040/

18 Chatham House, 2022, *Future without Plastic*, www.chathamhouse.org/2022/08/future-without-plastic

19 https://8billiontrees.com/carbon-offsets-credits/carbon-ecological-footprint-calculators/plastic-carbon-footprint/

20 Euractiv, 2023, *UN Plastic Treaty*, www.euractiv.com/section/circular-economy/news/un-plastic-treaty-talks-grapple-with-re-use-recycle-reduce-debate/

21 Common Dreams, 2023, *Global Plastics Treaty*, www.commondreams.org/opinion/biden-plastics-treaty-production

22 UNEP, 2018, *Alternative Materials*, www.unep.org/news-and-stories/press-release/exploring-alternative-materials-reduce-plastic-pollution

23 R.S. Dassanayake, 2018, *IntechOpen*, www.intechopen.com/chapters/63722

24 www.ncbi.nlm.nih.gov/pmc/articles/PMC9277587/

25 www.researchgate.net/figure/The-cycle-of-PLA-in-nature-While-PLA-can-be-considered-an-eco-friendly-biomaterial-with_fig1_221922647

26 Sustainable Plastics, 2021, *PHA: As Green as It Gets*, www.sustainableplastics.com/news/pha-green-it-gets

27 www.ncbi.nlm.nih.gov/pmc/articles/PMC9797907/

28 www.knowledge-share.eu/en/patent/recycling-of-plastics-from-biodegradable-bags/

29 Novamont, *Shopping Bags*, https://northamerica.novamont.com/leggi_evento.php?id_event=23

30 European Parliament, 2018, *Microplastics*, www.europarl.europa.eu/news/en/headlines/society/20181116STO19217/microplastics-sources-effects-and-solutions

31 V.N. Prapanchan, 2023, *Water*, Vol. 15, p. 1987, www.mdpi.com/2073-4441/15/11/1987

32 OECD, 2022, *Recycling*, www.oecd.org/environment/plastic-pollution-is-growing-relentlessly-as-waste-management-and-recycling-fall-short.htm

33 OECD, 2023, *Towards Eliminating Plastic Pollution by 2040*, www.oecd.org/environment/plastics/ and www.oecd.org/environment/plastics/Interim-Findings-Towards-Eliminating-Plastic-Pollution-by-2040-Policy-Scenario-Analysis.pdf

34 TNO, 2023, www.tno.nl/en/newsroom/2023/06/from-plastic-free-future-proof-plastics/

35 UNEP, 2024, *Intergovernmental Negotiating Committee*, www.unep.org/inc-plastic-pollution

36 European Commission, *Global Action on Plastics*, https://environment.ec.europa.eu/topics/plastics/global-action-plastics_en

37 UN News, 2023, *Climate and Environment*, https://news.un.org/en/interview/2023/11/1143727

38 J.-G. Rosenboom, 2022, *Nature Reviews Materials*, Vol. 7, p. 177, www.nature.com/articles/s41578-021-00407-8

39 J. Liu, www.frontiersin.org/articles/10.3389/fenvs.2023.1130463/full

Glossary

Acidogenesis: A biological reaction in which simple monomers are converted into volatile fatty acids; It is a biological reaction in which volatile fatty acids are converted into acetic acid, carbon dioxide, and hydrogen.

Acetogenesis: A substep of the acid-forming stage and is completed through carbohydrate fermentation, resulting in acetate, CO_2, and H_2 that methanogens can utilize to form methane.

Additives: Plastic is usually made from a polymer mixed with a complex compound of additives. These additives, which include flame retardants, plasticizers, pigments, fillers, and stabilizers, are used to improve the plastic's properties or to reduce its cost.

Alginate: An anionic polysaccharide mainly found in brown algae.

Amylopectin: The major component of starch by weight and is one of the largest molecules in nature.

Amylose: Constitutes 5–35% of most natural starches and influences starch properties in foods.

Anaerobic digestion: The process through which bacteria break down organic matter, involving hydrolysis, acidogenesis, acetogenesis, and methanogenesis.

Anthropogenic: Relating to or resulting from the influence of human beings on nature.

Biobased economy: Sometimes synonymous with bioeconomy, but more precisely a bioeconomy excluding food, feed, and primary biomass production.

Biobased plastic: Plastic derived fully or partially from plant materials, such as cellulose, potato or corn starch, sugar cane, maize, and soy, instead of petroleum or natural gas. They can be engineered to be biodegradable or compostable, but they can be designed to be structurally identical to petroleum-based plastics, in which case they can last in the environment for the same period of time.

Biobenign (materials): A material harmless to natural systems in case it unintentionally escapes collection and recovery systems.

Biocatalysis: Use of biological systems or their parts to catalyze chemical reactions.

Biodegradable (materials): A material that can, with the help of microorganisms, break down into natural components (e.g., water, carbon dioxide, or biomass) under certain conditions.

Biodegradation: The breakdown of organic matter by microorganisms, such as bacteria and fungi.

Bioeconomy: According to the EU definition, the bioeconomy encompasses the production of renewable biological resources and the conversion of these resources, residues, byproducts, and side streams into value-added products, such as food, feed, biobased products, services, and bioenergy.

Biofilm: A syntrophic community of microorganisms in which cells stick to each other and often also to a surface.

119

Biomass: Material of biological origin, excluding the material embedded in geologic formations and/or fossilized.

Business-as-usual scenario: Defined as a "no intervention" scenario; in other words, it assumes that the current policy framework, market dynamics, cultural norms, and consumer behaviors will not change.

CAGR: Compound annual growth rate. An investment's mean annual growth rate over a period longer than one year.

Carbon capture and utilization (CCU): Refers to a range of applications through which CO_2 is captured and used either directly (i.e., nonchemically altered) or indirectly (i.e., transformed) in various products. CO_2 is primarily used in the fertilizer industry and for enhanced oil recovery.

Carrageenans: A family of natural linear sulfated polysaccharides extracted from red edible seaweeds.

Cellulose: A polysaccharide composed of a linear chain of β-1,4–linked glucose units with a degree of polymerization ranging from several hundred to over 10,000, which is the most abundant organic polymer on Earth.

Chemical conversion: A process that breaks down polymers into individual monomers or other hydrocarbon products that can then serve as building blocks or feedstock to produce polymers again

Circular economy: One of the current sustainable economic models in which products and materials are designed in such a way that they can be reused, remanufactured, recycled, or recovered and thus maintained in the economy for as long as possible, along with the resources of which they are made, and the generation of waste, especially hazardous waste, is avoided or minimized, and greenhouse gas emissions are prevented or reduced, which can contribute significantly to sustainable consumption and production.

Chitin: The second most abundant polysaccharide in nature (behind only cellulose). It is a primary component of cell walls in fungi (especially filamentous and mushroom-forming fungi) and the exoskeletons of arthropods.

Chitosan: Made by treating the chitin shells of shrimp and other crustaceans with an alkaline substance, such as sodium hydroxide.

Circular plastic: Designed to be reused safely many times. Its material is recycled or composted at the end of use, in practice and at scale, minimizing its adverse environmental impacts and respecting the rights, health, and safety of all people involved across its life cycle, including product users.

Closed-loop recycling: The recycling of plastic into any new application that will eventually be found in municipal solid waste, essentially replacing virgin feedstock (i.e., plastic bottles, pens, etc.) (See "recycling.")

Compostable (materials): Materials, including compostable plastic and nonplastic materials, that are approved to meet local compostability standards (for example, industrial composting standard EN 13432, where industrial-equivalent composting is available).

Design for recycling: The principle and process by which companies design their products and packaging to be recyclable.

Downstream activities: These involve end-of-life management, including segregation, collection, sorting, recycling, and disposal. Recycling is a process that starts downstream and "closes the loop" by connecting upstream (i.e., starting a new life cycle for new plastic products with old materials). Similarly, repair/refurbishment processes provide another way to close the loop by bringing products back into the midstream.

Durability: Materials often selected for applications requiring resistance. It refers to plastics with average use cycles above three years, which are frequently used for industrial and construction applications.

Ecosystem: A system that environments and their organisms form through their interactions.

End-of-life (EOL): A generalized term used to describe the part of the life cycle following the use phase.

Enzymatic recycling: A process employing enzymes that were initially produced by bacteria to break down plastics.

Extended producer responsibility (EPR): An environmental policy approach in which a producer's responsibility for a product is extended to the waste stage of that product's life cycle. In practice, EPR involves producers taking responsibility for the management of products after they become waste, including collection; pretreatment (e.g., sorting, dismantling, or depollution); (preparation for) reuse; recovery (including recycling and energy recovery); and final disposal. EPR systems can allow producers to exercise their responsibility by providing the financial resources required and/or by taking over the operational aspects of the process from municipalities. They assume the responsibility voluntarily or mandatorily; EPR systems can be implemented individually or collectively.

Extrusion: Plastics extrusion is a high-volume manufacturing process in which raw plastic is melted and formed into a continuous profile.

Feedstock: Any bulk raw material that is the principal input for an industrial production process.

Fermentation: A metabolic process that consumes carbohydrates without oxygen through the action of enzymes produced by microorganisms like yeast and bacteria.

Fischer-Tropsch: A catalyzed chemical reaction in which synthesis gas is converted into liquid hydrocarbons of various forms.

Fouling: The accumulation of unwanted material on solid surfaces. The fouling materials can consist of either living organisms (biofouling, organic) or nonliving substances (inorganic).

Gellan gum: A water-soluble anionic polysaccharide produced by the bacterium *Sphingomonas elodea.*

"Ghost gear": Materials (like fishing nets) that have been abandoned, lost, or otherwise discarded in the ocean, lakes, and rivers.

Gum arabic: A natural gum originally consisting of the hardened sap of two species of the *Acacia* tree. However, "gum arabic" does not indicate a particular botanical source.

Gyre (oceanography): Any large system of circulating ocean surface currents.

Hemicelluloses: Several heteropolymers present along with cellulose in almost all terrestrial plant cell walls.

Hydrolysis: Any chemical reaction in which a molecule of water breaks one or more chemical bonds. The term is used broadly for substitution, elimination, and solvation reactions in which water is the nucleophile.

Incineration: A waste treatment process that involves the combustion of substances contained in waste materials.

Informal waste sector: Where workers and economic units are involved in solid waste collection, recovery, and recycling activities that are—in law or in practice—not covered or insufficiently covered by formal arrangements.

Intentionally added microplastics: Tiny plastic manufactured particles (usually of a size less than 5 mm) intentionally added to some products to perform a concrete function, like abrasion.

Leakage: Materials that do not follow an intended pathway and "escape" or are otherwise lost to the system. Litter is an example of system leakage.

Legacy plastic: Plastics that cannot be reused or recycled, including plastics that are already in the environment as existing pollution or are stocked or will enter the economy, for example, in short-lived or durable products designed without considering their circularity or long-term use in the economy.

Lignin: A class of complex organic polymers that form key structural materials in the support tissues of most plants.

Lignocellulose: Refers to plant dry matter, so-called lignocellulosic biomass. It is the most abundantly available raw material on Earth for the production of biofuels, mainly bio-ethanol.

Managed landfill: A place where collected waste has been deposited in a central location and where the waste is controlled through daily, intermediate, and final cover, thus preventing the top layer from escaping into the natural environment through wind and surface water.

Mechanical recycling: Processing of plastic waste into secondary raw material or products without significantly changing the material's chemical structure.

Microfibers: Microsize fragments (10 µm) to coarse fraction (>1 mm).

Microplastics: Fragments of any type of plastic less than 5 mm (0.20 in.) in length. They cause pollution by entering natural ecosystems from a variety of sources, including cosmetics, clothing, food packaging, and industrial processes.

Midstream activities: Involve the design, manufacture, packaging, distribution, use (and reuse), and maintenance of plastic products and services. Keeping plastic products at midstream as long as possible is ideal for circularity, because this is where plastic products have their highest value.

Mismanaged waste: Collected waste that has been released or deposited in a place from which it can move into the natural environment (intentionally or otherwise). This includes dumpsites and unmanaged landfills.

Nano-plastics: Less than 1 µm (i.e., 1000 nm) or less than 100 nm in size.

Net-zero emission: The state where emissions of greenhouse gases due to human activities and removal of these gases are in balance over a given period.

Organic recycling: A closed-loop activity where natural materials are converted from organic waste to energy, gas, water, and biomass.

Oxo-degradable: Products containing a pro-oxidant that induces the breakdown of the plastic product into smaller pieces under favorable conditions (e.g., heat, UV light, and mechanical stress).

The Paris Agreement: Also known as the Paris Accords or the Paris Climate Accords, this is an international climate change treaty adopted in 2015. It covers mitigation, adaptation, and finance.

Pathway: A course of action that combines system interventions across geographic archetypes to achieve a desired system outcome.

Plastic pellets: Microsize ($\leq$5 mm) granules, usually in the shape of a cylinder or a disk, produced as a raw material (also from plastic recycling) and used in the manufacture of plastic products.

Photo-oxidation: Oxidative photodegradation.

Plastic categories: Modeled as flowing separately through the system: rigid mono-material plastics, flexible mono-material plastics, multilayer plastics, and multimaterials:

- Rigid mono-material plastics—An item made from a single plastic polymer that holds its shape, such as a bottle or tub.
- Flexible mono-material plastics—An item made from a single plastic polymer that is thin, such as plastic wraps and bags.
- Multilayer plastics—An item, usually packaging, made of multiple plastic polymers that cannot be easily and mechanically separated.
- Multimaterials—An item, usually packaging, made of plastic and nonplastic materials (such as thin metal foils or cardboard layers) that cannot be easily and mechanically separated.

Plastic-to-fuel (P2F): The process by which the output material of chemical conversion plants is refined into alternative fuels such as diesel.

Plastic-to-plastic (P2P): Several chemical conversion technologies being developed that can produce petrochemical feedstock that can be reintroduced into the petrochemical process to produce virgin-like plastic.

Plastic pollution: The negative effects and emissions from producing and consuming plastic materials and products throughout their life cycle. This definition includes mismanaged plastic waste (e.g., open-burned and dumped in uncontrolled dumpsites) and leakage and accumulation of plastic objects and particles that can adversely affect humans and the living and nonliving environment.

Polymers:

- **Polycarbonates** are a group of thermoplastic polymers containing carbonate groups in their chemical structure.
- **Polyethylene succinate** (PES) is an aliphatic synthetic polyester.
- **Polyhydroxyalkanoate** (PHAS) are a family of intracellular microbial polyesters (PHB, PH3B, PHBV) synthesized by many species of Bacteria and Archaea.
- **Polyethylene terephthalate** (PET) is a polymer produced through the polymerization of ethylene glycol and terephthalic acid; it is mainly used in packaging and textile production applications.
- **Polyadipate terephthalate** (PBAT) is a biodegradable random copolymer. It is generally marketed as a fully biodegradable alternative to low-density

polyethylene, having many similar properties, including flexibility and resilience, allowing it to be used for many similar uses, such as plastic bags and wraps.

- **Polyethylene furanoate** (PEF) is a 100% recyclable, biobased polymer produced using renewable raw materials (sugars) from plants. It is an aromatic polyester made of ethylene glycol and is a chemical analogue of PET.
- **Polyethylene** (PE) is the most common type of consumer plastic. It is a thermoplastic product, meaning it can be melted into a liquid and then cooled back into a solid many times over. (HDPE—High-density polyethylene. LDPE—Low-density polyethylene, LLDPE—Linear low-density polyethylene)
- **Polypropylene** (PP) is a partially crystalline thermoplastic synthetic polymer belonging to the polyolefin group. It is used in a variety of applications.
- **Polyvinyl chloride** (PVC) is the world's third-most widely produced synthetic polymer of plastic. About 40 million tons of PVC are produced each year. It comes in rigid and flexible forms. Rigid PVC is used in the construction of pipes, doors, and windows.
- **Polystyrene** (PS) is a synthetic polymer made from monomers of the aromatic hydrocarbon styrene. Expanded polystyrene can be solid or foamed.
- **Polyamide** 6 (PA6) is a thermoplastic polymer of the nylon family.
- **Polycaprolactone** (PCL) is a synthetic, semi-crystalline, biodegradable polyester.
- **Polylactic acid** (PLA) is a thermoplastic polyester from a monomer made from fermented plant starch such as corn, cassava, sugarcane, or sugar beet pulp.
- **Polyvinyl alcohol** (PVOH, PVA, or PVAl) is a water-soluble synthetic polymer. It is used in papermaking, textile warp sizing, as a thickener and emulsion stabilizer in polyvinyl acetate (PVAc) adhesive formulations, in a variety of coatings, and in 3D printing.

Primary microplastics: Microplastics are fragments of any type of plastic less than 5 mm (0.20 in.) in length. They cause pollution by entering natural ecosystems from a variety of sources, including cosmetics, clothing, food packaging, and industrial processes.

Pyrolysis: Thermochemical decomposition of organic material at elevated temperature (>430 °C) in the absence of oxygen.

Recyclable: For something to be deemed recyclable, the system must be in place for it to be collected, sorted, reprocessed, and manufactured back into a new product or packaging—at scale and economically. Recyclable is used here as a shorthand for "mechanically recyclable." (See "Mechanical recycling.")

Recycling: Processing waste materials for the original purpose or other purposes, excluding energy recovery.

Renewable: A renewable resource is a substance of economic value that is replenished naturally over time, thereby supporting sustainability despite consumption.

Reusable: Products and packaging, including plastic bags, that are conceived and designed to accomplish a minimum number of uses within their life cycle for the same purpose for which they were conceived. In terms of "minimum number of uses," the PR3 Standards suggest that reusable (containers) should be designed to withstand at least ten reuse cycles.

Reuse: Using a product more than once in its original form.

Reuse, recycle, and reorient and diversify: Three necessary market shifts that emphasize the need to eliminate unnecessary and problematic plastic uses.

Ring-opening polymerization: A form of chain-growth polymerization in which the terminus of a polymer chain attacks cyclic monomers to form a longer polymer

Safe disposal: Ensuring that any waste that reaches its end-of-life is disposed of in a way that does not cause leakage of plastic waste or chemicals into the environment, does not pose hazardous risks to human health, and, in the case of landfills, is contained securely for the long term.

Secondary microplastics: Small particle pieces that have resulted from the fragmentation and weathering of larger plastic items.

Short-lived plastic products: Plastics within the packaging and consumer products. These are the two categories of plastic products with the shortest average use cycles—0.5 and 3 years, respectively. Note that the categorization is based on average life span; therefore, some products in this category will, in practice, have longer life spans than three years.

Solvolysis: A type of nucleophilic substitution or elimination where the nucleophile is a solvent molecule.

Starch (polysaccharides): Produced by most green plants for energy storage. Thermoplastic starch (TPS) is produced by mixing native starch with a plasticizer at a temperature above the starch gelatinization temperature, typically in the 70–90 °C range.

Synthetic rubber: An artificial elastomer synthesized from petroleum byproducts.

Systems change: Captures the idea of addressing the causes rather than the symptoms of a societal issue by taking a holistic scenario.

Thermochemical recycling: A process based on the thermal decomposition of synthetic organic substances (especially plastics) by the action of heat without access to oxidizing substances.

Thermoplastics or thermo-softening plastic: Any plastic polymer material that becomes pliable or moldable at a certain elevated temperature and solidifies upon cooling.

Tire dust: Consists of micro-size particles with a spectrum from airborne (>10 μm) to a coarse fraction (>1 mm) released through mechanical abrasion of tires, with chemical composition depending on rubber type.

Upstream activities: These include obtaining the raw materials from crude oil, natural gas, or recycled and renewable feedstock (e.g., biomass) and polymerization. Plastic leakage into the environment (e.g., pellets and flakes) already happens at this stage.

Waste management: The processes and actions required to manage waste from its inception to its final disposal.

Index

Note: numbers in **bold** indicate a table. Numbers in *italics* indicate a figure on the corresponding page.

Printed in the United States
by Baker & Taylor Publisher Services